£21.00

Pitman Research Notes in Mathematics Series

AF838104

Submission of proposals for consideration
Suggestions for publication, in the form of outlines and representative samples, are invited by the Editorial Board for assessment. Intending authors should approach one of the main editors or another member of the Editorial Board, citing the relevant AMS subject classifications. Alternatively, outlines may be sent directly to the publisher's offices. Refereeing is by members of the board and other mathematical authorities in the topic concerned, throughout the world.

Preparation of accepted manuscripts
On acceptance of a proposal, the publisher will supply full instructions for the preparation of manuscripts in a form suitable for direct photo-lithographic reproduction. Specially printed grid sheets are provided and a contribution is offered by the publisher towards the cost of typing. Word processor output, subject to the publisher's approval, is also acceptable.

Illustrations should be prepared by the authors, ready for direct reproduction without further improvement. The use of hand-drawn symbols should be avoided wherever possible, in order to maintain maximum clarity of the text.

The publisher will be pleased to give any guidance necessary during the preparation of a typescript, and will be happy to answer any queries.

Important note
In order to avoid later retyping, intending authors are strongly urged not to begin final preparation of a typescript before receiving the publisher's guidelines and special paper. In this way it is hoped to preserve the uniform appearance of the series.

Longman Scientific & Technical
Longman House
Burnt Mill
Harlow, Essex, UK
(tel (0279) 426721)

Titles in this series

1 Improperly posed boundary value problems
A Carasso and A P Stone
2 Lie algebras generated by finite dimensional ideals
I N Stewart
3 Bifurcation problems in nonlinear elasticity
R W Dickey
4 Partial differential equations in the complex domain
D L Colton
5 Quasilinear hyperbolic systems and waves
A Jeffrey
6 Solution of boundary value problems by the method of integral operators
D L Colton
7 Taylor expansions and catastrophes
T Poston and I N Stewart
8 Function theoretic methods in differential equations
R P Gilbert and R J Weinacht
9 Differential topology with a view to applications
D R J Chillingworth
10 Characteristic classes of foliations
H V Pittie
11 Stochastic integration and generalized martingales
A U Kussmaul
12 Zeta-functions: An introduction to algebraic geometry
A D Thomas
13 Explicit *a priori* inequalities with applications to boundary value problems
V G Sigillito
14 Nonlinear diffusion
W E Fitzgibbon III and H F Walker
15 Unsolved problems concerning lattice points
J Hammer
16 Edge-colourings of graphs
S Fiorini and R J Wilson
17 Nonlinear analysis and mechanics: Heriot-Watt Symposium Volume I
R J Knops
18 Actions of fine abelian groups
C Kosniowski
19 Closed graph theorems and webbed spaces
M De Wilde
20 Singular perturbation techniques applied to integro-differential equations
H Grabmüller
21 Retarded functional differential equations: A global point of view
S E A Mohammed
22 Multiparameter spectral theory in Hilbert space
B D Sleeman
24 Mathematical modelling techniques
R Aris
25 Singular points of smooth mappings
C G Gibson
26 Nonlinear evolution equations solvable by the spectral transform
F Calogero
27 Nonlinear analysis and mechanics: Heriot-Watt Symposium Volume II
R J Knops
28 Constructive functional analysis
D S Bridges
29 Elongational flows: Aspects of the behaviour of model elasticoviscous fluids
C J S Petrie
30 Nonlinear analysis and mechanics: Heriot-Watt Symposium Volume III
R J Knops
31 Fractional calculus and integral transforms of generalized functions
A C McBride
32 Complex manifold techniques in theoretical physics
D E Lerner and P D Sommers
33 Hilbert's third problem: scissors congruence
C-H Sah
34 Graph theory and combinatorics
R J Wilson
35 The Tricomi equation with applications to the theory of plane transonic flow
A R Manwell
36 Abstract differential equations
S D Zaidman
37 Advances in twistor theory
L P Hughston and R S Ward
38 Operator theory and functional analysis
I Erdelyi
39 Nonlinear analysis and mechanics: Heriot-Watt Symposium Volume IV
R J Knops
40 Singular systems of differential equations
S L Campbell
41 N-dimensional crystallography
R L E Schwarzenberger
42 Nonlinear partial differential equations in physical problems
D Graffi
43 Shifts and periodicity for right invertible operators
D Przeworska-Rolewicz
44 Rings with chain conditions
A W Chatters and C R Hajarnavis
45 Moduli, deformations and classifications of compact complex manifolds
D Sundararaman
46 Nonlinear problems of analysis in geometry and mechanics
M Atteia, D Bancel and I Gumowski
47 Algorithmic methods in optimal control
W A Gruver and E Sachs
48 Abstract Cauchy problems and functional differential equations
F Kappel and W Schappacher
49 Sequence spaces
W H Ruckle
50 Recent contributions to nonlinear partial differential equations
H Berestycki and H Brezis
51 Subnormal operators
J B Conway

52 Wave propagation in viscoelastic media
F Mainardi
53 Nonlinear partial differential equations and their applications: Collège de France Seminar. Volume I
H Brezis and J L Lions
54 Geometry of Coxeter groups
H Hiller
55 Cusps of Gauss mappings
T Banchoff, T Gaffney and C McCrory
56 An approach to algebraic K-theory
A J Berrick
57 Convex analysis and optimization
J-P Aubin and R B Vintner
58 Convex analysis with applications in the differentiation of convex functions
J R Giles
59 Weak and variational methods for moving boundary problems
C M Elliott and J R Ockendon
60 Nonlinear partial differential equations and their applications: Collège de France Seminar. Volume II
H Brezis and J L Lions
61 Singular systems of differential equations II
S L Campbell
62 Rates of convergence in the central limit theorem
Peter Hall
63 Solution of differential equations by means of one-parameter groups
J M Hill
64 Hankel operators on Hilbert space
S C Power
65 Schrödinger-type operators with continuous spectra
M S P Eastham and H Kalf
66 Recent applications of generalized inverses
S L Campbell
67 Riesz and Fredholm theory in Banach algebra
B A Barnes, G J Murphy, M R F Smyth and T T West
68 Evolution equations and their applications
F Kappel and W Schappacher
69 Generalized solutions of Hamilton-Jacobi equations
P L Lions
70 Nonlinear partial differential equations and their applications: Collège de France Seminar. Volume III
H Brezis and J L Lions
71 Spectral theory and wave operators for the Schrödinger equation
A M Berthier
72 Approximation of Hilbert space operators I
D A Herrero
73 Vector valued Nevanlinna Theory
H J W Ziegler
74 Instability, nonexistence and weighted energy methods in fluid dynamics and related theories
B Straughan
75 Local bifurcation and symmetry
A Vanderbauwhede
76 Clifford analysis
F Brackx, R Delanghe and F Sommen
77 Nonlinear equivalence, reduction of PDEs to ODEs and fast convergent numerical methods
E E Rosinger
78 Free boundary problems, theory and applications. Volume I
A Fasano and M Primicerio
79 Free boundary problems, theory and applications. Volume II
A Fasano and M Primicerio
80 Symplectic geometry
A Crumeyrolle and J Grifone
81 An algorithmic analysis of a communication model with retransmission of flawed messages
D M Lucantoni
82 Geometric games and their applications
W H Ruckle
83 Additive groups of rings
S Feigelstock
84 Nonlinear partial differential equations and their applications: Collège de France Seminar. Volume IV
H Brezis and J L Lions
85 Multiplicative functionals on topological algebras
T Husain
86 Hamilton-Jacobi equations in Hilbert spaces
V Barbu and G Da Prato
87 Harmonic maps with symmetry, harmonic morphisms and deformations of metrics
P Baird
88 Similarity solutions of nonlinear partial differential equations
L Dresner
89 Contributions to nonlinear partial differential equations
C Bardos, A Damlamian, J I Díaz and J Hernández
90 Banach and Hilbert spaces of vector-valued functions
J Burbea and P Masani
91 Control and observation of neutral systems
D Salamon
92 Banach bundles, Banach modules and automorphisms of C*-algebras
M J Dupré and R M Gillette
93 Nonlinear partial differential equations and their applications: Collège de France Seminar. Volume V
H Brezis and J L Lions
94 Computer algebra in applied mathematics: an introduction to MACSYMA
R H Rand
95 Advances in nonlinear waves. Volume I
L Debnath
96 FC-groups
M J Tomkinson
97 Topics in relaxation and ellipsoidal methods
M Akgül
98 Analogue of the group algebra for topological semigroups
H Dzinotyiweyi
99 Stochastic functional differential equations
S E A Mohammed

100 Optimal control of variational inequalities
V Barbu
101 Partial differential equations and dynamical systems
W E Fitzgibbon III
102 Approximation of Hilbert space operators. Volume II
C Apostol, L A Fialkow, D A Herrero and D Voiculescu
103 Nondiscrete induction and iterative processes
V Ptak and F-A Potra
104 Analytic functions – growth aspects
O P Juneja and G P Kapoor
105 Theory of Tikhonov regularization for Fredholm equations of the first kind
C W Groetsch
106 Nonlinear partial differential equations and free boundaries. Volume I
J I Díaz
107 Tight and taut immersions of manifolds
T E Cecil and P J Ryan
108 A layering method for viscous, incompressible L_p flows occupying R^n
A Douglis and E B Fabes
109 Nonlinear partial differential equations and their applications: Collège de France Seminar. Volume VI
H Brezis and J L Lions
110 Finite generalized quadrangles
S E Payne and J A Thas
111 Advances in nonlinear waves. Volume II
L Debnath
112 Topics in several complex variables
E Ramírez de Arellano and D Sundararaman
113 Differential equations, flow invariance and applications
N H Pavel
114 Geometrical combinatorics
F C Holroyd and R J Wilson
115 Generators of strongly continuous semigroups
J A van Casteren
116 Growth of algebras and Gelfand–Kirillov dimension
G R Krause and T H Lenagan
117 Theory of bases and cones
P K Kamthan and M Gupta
118 Linear groups and permutations
A R Camina and E A Whelan
119 General Wiener–Hopf factorization methods
F-O Speck
120 Free boundary problems: applications and theory, Volume III
A Bossavit, A Damlamian and M Fremond
121 Free boundary problems: applications and theory, Volume IV
A Bossavit, A Damlamian and M Fremond
122 Nonlinear partial differential equations and their applications: Collège de France Seminar. Volume VII
H Brezis and J L Lions
123 Geometric methods in operator algebras
H Araki and E G Effros
124 Infinite dimensional analysis–stochastic processes
S Albeverio
125 Ennio de Giorgi Colloquium
P Krée
126 Almost-periodic functions in abstract spaces
S Zaidman
127 Nonlinear variational problems
A Marino, L Modica, S Spagnolo and M Degiovanni
128 Second-order systems of partial differential equations in the plane
L K Hua, W Lin and C-Q Wu
129 Asymptotics of high-order ordinary differential equations
R B Paris and A D Wood
130 Stochastic differential equations
R Wu
131 Differential geometry
L A Cordero
132 Nonlinear differential equations
J K Hale and P Martinez-Amores
133 Approximation theory and applications
S P Singh
134 Near-rings and their links with groups
J D P Meldrum
135 Estimating eigenvalues with *a posteriori/a priori* inequalities
J R Kuttler and V G Sigillito
136 Regular semigroups as extensions
F J Pastijn and M Petrich
137 Representations of rank one Lie groups
D H Collingwood
138 Fractional calculus
G F Roach and A C McBride
139 Hamilton's principle in continuum mechanics
A Bedford
140 Numerical analysis
D F Griffiths and G A Watson
141 Semigroups, theory and applications. Volume I
H Brezis, M G Crandall and F Kappel
142 Distribution theorems of L-functions
D Joyner
143 Recent developments in structured continua
D De Kee and P Kaloni
144 Functional analysis and two-point differential operators
J Locker
145 Numerical methods for partial differential equations
S I Hariharan and T H Moulden
146 Completely bounded maps and dilations
V I Paulsen
147 Harmonic analysis on the Heisenberg nilpotent Lie group
W Schempp
148 Contributions to modern calculus of variations
L Cesari
149 Nonlinear parabolic equations: qualitative properties of solutions
L Boccardo and A Tesei
150 From local times to global geometry, control and physics
K D Elworthy

151 A stochastic maximum principle for optimal control of diffusions
U G Haussmann
152 Semigroups, theory and applications. Volume II
H Brezis, M G Crandall and F Kappel
153 A general theory of integration in function spaces
P Muldowney
154 Oakland Conference on partial differential equations and applied mathematics
L R Bragg and J W Dettman
155 Contributions to nonlinear partial differential equations. Volume II
J I Díaz and P L Lions
156 Semigroups of linear operators: an introduction
A C McBride
157 Ordinary and partial differential equations
B D Sleeman and R J Jarvis
158 Hyperbolic equations
F Colombini and M K V Murthy
159 Linear topologies on a ring: an overview
J S Golan
160 Dynamical systems and bifurcation theory
M I Camacho, M J Pacifico and F Takens
161 Branched coverings and algebraic functions
M Namba
162 Perturbation bounds for matrix eigenvalues
R Bhatia
163 Defect minimization in operator equations: theory and applications
R Reemtsen
164 Multidimensional Brownian excursions and potential theory
K Burdzy
165 Viscosity solutions and optimal control
R J Elliott
166 Nonlinear partial differential equations and their applications. Collège de France Seminar. Volume VIII
H Brezis and J L Lions
167 Theory and applications of inverse problems
H Haario
168 Energy stability and convection
G P Galdi and B Straughan
169 Additive groups of rings. Volume II
S Feigelstock
170 Numerical analysis 1987
D F Griffiths and G A Watson
171 Surveys of some recent results in operator theory. Volume I
J B Conway and B B Morrel
172 Amenable Banach algebras
J-P Pier
173 Pseudo-orbits of contact forms
A Bahri
174 Poisson algebras and Poisson manifolds
K H Bhaskara and K Viswanath
175 Maximum principles and eigenvalue problems in partial differential equations
P W Schaefer
176 Mathematical analysis of nonlinear, dynamic processes
K U Grusa
177 Cordes' two-parameter spectral representation theory
D F McGhee and R H Picard
178 Equivariant K-theory for proper actions
N C Phillips
179 Elliptic operators, topology and asymptotic methods
J Roe
180 Nonlinear evolution equations
J K Engelbrecht, V E Fridman and E N Pelinovski
181 Nonlinear partial differential equations and their applications. Collège de France Seminar. Volume IX
H Brezis and J L Lions
182 Critical points at infinity in some variational problems
A Bahri
183 Recent developments in hyperbolic equations
L Cattabriga, F Colombini, M K V Murthy and S Spagnolo
184 Optimization and identification of systems governed by evolution equations on Banach space
N U Ahmed
185 Free boundary problems: theory and applications. Volume I
K H Hoffmann and J Sprekels
186 Free boundary problems: theory and applications. Volume II
K H Hoffmann and J Sprekels
187 An introduction to intersection homology theory
F Kirwan
188 Derivatives, nuclei and dimensions on the frame of torsion theories
J S Golan and H Simmons
189 Theory of reproducing kernels and its applications
S Saitoh
190 Volterra integrodifferential equations in Banach spaces and applications
G Da Prato and M Iannelli
191 Nest algebras
K R Davidson
192 Surveys of some recent results in operator theory. Volume II
J B Conway and B B Morrel
193 Nonlinear variational problems. Volume II
A Marino and M K Murthy
194 Stochastic processes with multidimensional parameter
M E Dozzi
195 Prestressed bodies
D Iesan
196 Hilbert space approach to some classical transforms
R H Picard
197 Stochastic calculus in application
J R Norris
198 Radical theory
B J Gardner
199 The C* – algebras of a class of solvable Lie groups
X Wang

200 Stochastic analysis, path integration and dynamics
D Elworthy
201 Riemannian geometry and holonomy groups
S Salamon
202 Strong asymptotics for extremal errors and polynomials associated with Erdös type weights
D S Lubinsky
203 Optimal control of diffusion processes
V S Borkar
204 Rings, modules and radicals
B J Gardner
205 Two-parameter eigenvalue problems in ordinary differential equations
M Faierman
206 Distributions and analytic functions
R D Carmichael and D Mitrović
207 Semicontinuity, relaxation and integral representation in the calculus of variations
G Buttazzo
208 Recent advances in nonlinear elliptic and parabolic problems
P Bénilan, M Chipot, L Evans and M Pierre
209 Model completions, ring representations and the topology of the Pierce sheaf
A Carson
210 Retarded dynamical systems
G Stepan
211 Function spaces, differential operators and nonlinear analysis
L Paivarinta
212 Analytic function theory of one complex variable
C C Yang, Y Komatu and K Niino
213 Elements of stability of visco-elastic fluids
J Dunwoody
214 Jordan decompositions of generalised vector measures
K D Schmidt
215 A mathematical analysis of bending of plates with transverse shear deformation
C Constanda
216 Ordinary and partial differential equations Vol II
B D Sleeman and R J Jarvis
217 Hilbert modules over function algebras
R G Douglas and V I Paulsen
218 Graph colourings
R Wilson and R Nelson
219 Hardy-type inequalities
A Kufner and B Opic
220 Nonlinear partial differential equations and their applications. College de France Seminar Volume X
H Brezis and J L Lions
221 Workshop on dynamical systems
E Shiels and Z Coelho
222 Geometry and analysis in nonlinear dynamics
H W Broer and F Takens
223 Fluid dynamical aspects of combustion theory
M Onofri and A Tesei
224 Approximation of Hilbert space operators. Volume I. 2nd edition
D Herrero
225 Operator Theory: Proceedings of the 1988 GPOTS–Wabash conference
J B Conway and B B Morrel
226 Local cohomology and localization
J L Bueso Montero, B Torrecillas Jover and A Verschoren
227 Nonlinear waves and dissipative effects
D Fusco and A Jeffrey
228 Numerical analysis. Volume III
D F Griffiths and G A Watson
229 Recent developments in structured continua. Volume III
D De Kee and P Kaloni
230 Boolean methods in interpolation and approximation
F J Delvos and W Schempp
231 Further advances in twistor theory, Volume 1
L J Mason and L P Hughston
232 Further advances in twistor theory, Volume 2
L J Mason and L P Hughston
233 Geometry in the neighborhood of invariant manifolds of maps and flows and linearization
U Kirchgraber and K Palmer
234 Quantales and their applications
K I Rosenthal
235 Integral equations and inverse problems
V Petkov and R Lazarov
236 Pseudo-differential operators
S R Simanca
237 A functional analytic approach to statistical experiments
I M Bomze
238 Quantum mechanics, algebras and distributions
D Dubin and M Hennings
239 Hamilton flows and evolution semigroups
J Gzyl
240 Topics in controlled Markov chains
V S Borkar
241 Invariant manifold theory for hydrodynamic transition
S Sritharan
242 Lectures on the spectrum of $L^2(\Gamma\backslash G)$
F L Williams

M Chipot and
J Saint Jean Paulin (Editors)
University of Metz, France

Progress in partial differential equations: the Metz surveys

Longman
Scientific &
Technical
Copublished in the United States with
John Wiley & Sons, Inc., New York

Longman Scientific & Technical,
Longman Group UK Limited,
Longman House, Burnt Mill, Harlow,
Essex CM20 2JE, England
and Associated Companies throughout the world.

Copublished in the United States with
John Wiley & Sons, Inc., 605 Third Avenue, New York, NY 10158

First published 1991

AMS Subject Classification: (Main) 35X, 65XX
(Subsidiary) 49A29

ISSN 0269-3674

British Library Cataloguing in Publication Data
Progress in partial differential equations: The Metz surveys.
I. Chipot, M. II. Paulin, J. Saint Jean
515

ISBN 0-582-07024-4

Library of Congress Cataloging-in-Publication Data
is available

Printed and bound in Great Britain
by Biddles Ltd, Guildford and King's Lynn

CONTENTS

1. Calculus of Variations

2. Hyperbolic Problems - Exact Controllability

3. Numerical Analysis

PREFACE

This volume collects part of the conferences given at the University of Metz during the 1989-1990 "Metz Days".

Three important topics in Partial Differential Equations are concerned : Calculus of Variations, Hyperbolic Problems and Exact Controllability, Numerical Analysis.

Numerical analysis has been recently an important factor of progress in new developments of partial differential equations. This is especially true in Calculus of Variations and Control Theory.

The authors have attempted to write their contribution keeping in mind two goals: present their last results and give a short introduction to their area of research. In particular they have tried to introduce the reader to their work through some simple cases. We hope that this attempt will help the mathematical and more generally the scientific community in spreading quickly important and new techniques.

We would like to thank the city of Metz, the "Région de Lorraine" and the University of Metz for their support in organizing these conferences.

Finally, we express our special thanks to Longman for helping us to publish these proceedings.

M. Chipot and J. Saint Jean Paulin

CALCULUS OF VARIATIONS

B DACOROGNA

Existence and regularity of diffeomorphism with prescribed Jacobian

Département de Mathématiques
Ecole Polytechnique Fédérale de Lausanne
1015 LAUSANNE, Switzerland

I. INTRODUCTION:

In this article we are concerned with the following problem, find a diffeomorphism $u: \bar{\Omega} \to \bar{\Omega}$, where $\Omega \subset \mathbb{R}^n$ is a bounded open set, satisfying

$$\begin{cases} g(u(x))\det\nabla u(x) = f(x) & x \in \Omega \\ u(x) = x & x \in \partial\Omega \end{cases} \tag{1.1}$$

where $f,g > 0$ in $\bar{\Omega}$ and satisfy the compatibility condition

$$\int_\Omega g(x)dx = \int_\Omega f(x)dx. \tag{1.2}$$

In geometrical terms the problem is to show that two volume elements of the some total volume (i.e.(1.2)) are diffeomorphically equivalent.

Let us state precisely the result.

Theorem 1:

Let $k \geq 0$ be an integer, $0 < \alpha < 1$ and $\Omega \subset \mathbb{R}^n$ a bounded open set with $C^{k+3,\alpha}$ boundary $\partial\Omega$ ($C^{k,\alpha}$ denoting the usual Hölder space). Let $f,g \in C^{k,\alpha}(\bar{\Omega})$, $f,g > 0$ in $\bar{\Omega}$ satisfying (1.2). Then there exists $u \in \mathrm{Diff}^{k+1,\alpha}(\bar{\Omega})$ (the set of diffeomorphisms $u: \bar{\Omega} \to \bar{\Omega}$ with $u,u^{-1} \in C^{k+1,\alpha}(\bar{\Omega},\bar{\Omega})$) satisfying (1.1).

Remark:

It is clear that (1.1) is underdetermined and uniqueness of solutions does not hold. Indeed let $f = g = 1$, N an integer then $u(x) = r(\cos(\theta+2N\pi r^2), \sin(\theta+2N\pi r^2))$ satisfy (1.1) when Ω is the unit disk of $\mathbb{R}^2$ and (r,θ) are the polar coordinates.

The above result has been obtained by Dacorogna-Moser [1]. The first result concerning (1.1) was established by Moser [1] for manifolds without boundary and for $k = \infty$. It was then extended by Banyaga [1] for manifolds with boundary and for $k = \infty$. Zehnder [1] obtained Theorem 1 for manifolds without boundary and with a smallness assumption on the $C^{0,\alpha}$ norm of $(f-g)$. Independently Tartar [1] and Dacorogna [1] proved this result for $k = \infty$ and Ω diffeomorphic to the unit ball of $\mathbb{R}^2$ and $\mathbb{R}^3$ respectively.

In Dacorogna-Moser [1] it is also shown how to weaken the regularity of the boundary $\partial\Omega$, so as to include Lipschitz boundaries as well as isolated boundary points (as in a punctured disk)... In this article it is shown that if $k > 0$, $f,g \in C^k(\Omega)$ ($f,g > 0$ in $\bar{\Omega}$, $f + \frac{1}{f}$ and $g + \frac{1}{g}$ bounded) and satisfy (1.2), then there exists a diffeomorphism $u \in \text{Diff}^k(\Omega) \cap \text{Diff}^o(\bar{\Omega})$ (where $\text{Diff}^o(\bar{\Omega})$ stands for the set of homemorphisms $u\colon \bar{\Omega} \to \bar{\Omega}$ with $u, u^{-1} \in C^o(\bar{\Omega};\bar{\Omega})$). If $k = 0$ and f,g are as above, it is proved that there exists $u \in \text{Diff}^o(\bar{\Omega})$ satisfying a weak form of (1.1) namely

$$\int_E f(x)dx = \int_{u(E)} g(x)dx \tag{1.3}$$

for every open set $E \subset \Omega$ and $u(x) = x$ on $\partial\Omega$. Note that this result is weaker, from the point of view of regularity, than Theorem 1, since the expected gain in regularity is not ensured.

Theorem 1 has several applications. The first one is that we can construct a volume preserving diffeomorphism with given boundary data. More precisely let $\psi_o \in \text{Diff}^{k,\alpha}(\bar{\Omega})$ with $k \geq 1$, then there exists $\psi \in \text{Diff}^{k,\alpha}(\bar{\Omega})$ satisfying

$$\begin{cases} \det\nabla\psi \equiv 1 & \text{in } \Omega \\ \psi = \psi_o & \text{on } \partial\Omega. \end{cases} \tag{1.4}$$

With the above hypotheses (1.4) is equivalent to (1.1) and the existence is therefore deduced from Theorem 1.

The construction of volume preserving diffeomorphism plays a role in ergodic theory (c.f. Alpern [1], Anosov-Katok [1]) and is important in incompressible elasticity (c.f. Ball [1], Dacorogna [2]). It can also be used in the calculus of variations and in problems of equilibrium of gases (c.f. Dacorogna [1],[2]). For example if one looks for solutions of

$$\text{(P)} \qquad \inf\{\int_\Omega \phi(\det\nabla u(x))dx : u = u_0 \text{ on } \partial\Omega,\ u \in \mathrm{Diff}^{k,\alpha}(\bar{\Omega})\}$$

where $\phi: \mathbb{R} \to \mathbb{R}$ is convex and $u_0 \in \mathrm{Diff}^{k,\alpha}(\bar{\Omega})$ with $k \geqq 1$. Indeed it follows from Jensen inequality that

$$\int_\Omega \phi(\det\nabla u(x))dx$$

$$\geqq \mathrm{meas}\Omega\ \phi\left(\frac{1}{\mathrm{meas}\Omega}\int_\Omega \det\nabla u(x)dx\right) = \mathrm{meas}\Omega\ \phi\left(\frac{1}{\mathrm{meas}\Omega}\int_\Omega \det\nabla u_0(x)dx\right)$$

$$\mathrm{meas}\Omega\ \phi(1);$$

the last indentity following from the fact that $u_0(\Omega) = \Omega$. Therefore using the fact that there exists $\bar{u} \in \mathrm{Diff}^{k,\alpha}(\bar{\Omega})$ satisfying (1.4) we find, because of the above inequality, that $\bar{u}$ is actually a minimum of (P).

Before concluding with some open problems, let us mention an interesting result for the corresponding linear problem, which will be used for obtaining Theorem 1.

<u>Theorem 2</u>:

Let $k \geqq 0$ be an integer, $0 < \alpha < 1$ and $\Omega \subset \mathbb{R}^n$ a bounded open set with $C^{k+3,\alpha}$ boundary $\partial\Omega$. Let $f \in C^{k,\alpha}(\bar{\Omega})$ with

$$\int_\Omega f(x)dx = 0. \qquad (1.5)$$

Then there exists $u \in C^{k+1,\alpha}(\bar{\Omega};\mathbb{R}^n)$ satisfying

$$\begin{cases} \operatorname{div} u = f & \text{in } \Omega \\ u = 0 & \text{on } \partial\Omega . \end{cases} \tag{1.6}$$

Remark:

Several authors have proved Theorem 2 under some slightly different conditions, however as stated above it has been established by Dacorogna-Moser [1].

We now state three open problems

Problem 1:

Let Ω be C^∞ smooth, $k \geq 0$, $f \in C^k(\bar{\Omega})$, $f > 0$ in $\bar{\Omega}$ satisfying (1.2) with $g \equiv 1$. Does there exist $u \in \mathrm{Diff}^{k+1}(\bar{\Omega})$ satisfying (1.1)?

Problem 2:

Let Ω be C^∞ smooth, $k \geq 1$, $f \in C^k(\bar{\Omega})$ satisfying (1.5). Does there exist $u \in C^{k+1}(\bar{\Omega}; \mathbb{R}^n)$ satisfying (1.6)?

Problem 3:

Let Ω be C^∞-smooth, $k \geq 0$, $p > 1$, $f \in W^{k,p}(\Omega)$ (the usual Sobolev space), $f(x) \geq C_0 > 0$ for almost all $x \in \Omega$ and f satisfying (1.2) with $g \equiv 1$. Does there exist $u \in W^{k+1,np}(\Omega)$ satisfying (1.1)?

Remarks

i) Note that C^k as well as $C^{k,\alpha}$ spaces have the advantage over L^p spaces to form an algebra. In that sense one has, when working with Sobolev spaces with low k and p, to ask for solution in $W^{k+1,np}$ instead of $W^{k+1,p}$ (this last case can be treated by the method developed below).

ii) The third conjecture is false when k = 0 and p = 1. Indeed Müller [1] and Coifman-Lions-Meyer-Semmes [1] have shown that given any $u \in W^{1,n}(\Omega)$ with $\det\nabla u \geq 0$ a.e. then $\det\nabla u \log(2+\det\nabla u) \in L^1_{loc}(\Omega)$.

II. PROOF OF THEOREM 1 AND 2

We now sketch the proofs of the theorems and we refer for more details to Dacorogna-Moser [1]. We first assume that Theorem 2 has been proved.

Proof of Theorem 1:

The proof is divided in four steps.

Step 1: We first show that there is no loss of generality in assuming $g \equiv 1$ in (1.1). Indeed writing

$$\begin{cases} \det\nabla\phi(x) = \dfrac{\text{meas}\,\Omega}{\int_\Omega f(x)dx} f(x) & \text{in } \Omega \\ \det\nabla\psi(x) = \dfrac{\text{meas}\,\Omega}{\int_\Omega g(x)dx} g(x) & \text{in } \Omega \\ \phi(x) = \psi(x) = x & \text{on } \partial\Omega \end{cases} \tag{2.1}$$

we have that $u = \psi^{-1} \circ \phi$ satisfies (1.1).

Step 2: We then obtain Theorem 1 without the regularity result, i.e. we find $u \in \text{Diff}^{k,\alpha}(\bar{\Omega})$ (instead of $\text{Diff}^{k+1,\alpha}$) satisfying (1.1) with $g \equiv 1$. This is obtained by a deformation argument explained below.

Step 3: Using Theorem 2 and a smallness assumption on the $C^{0,\beta}$ norm $0 < \beta < \alpha < 1$ of f-1 we obtain by a fixed point argument a $\mathrm{Diff}^{k+1,\alpha}$ solution of (1.1).

Step 4: We remove the smallness assumption in Step 3 by combining two diffeomorphisms one constructed via Step 2 the other one via Step 3.

We now prove Step 2.

Step 2: Let $v \in C^{k+1,\alpha}(\bar{\Omega};\mathbb{R}^n)$ satisfying

$$\begin{cases} \operatorname{div} v = f-1 & \text{in } \Omega \\ v = 0 & \text{on } \partial\Omega. \end{cases} \tag{2.2}$$

Such a v exists by Theorem 2 and the fact that $g \equiv 1$ in (1.2) i.e.

$$\int_\Omega (f-1)dx = 0.$$

Assume that $k \geqq 1$. Define then a vector field

$$v(t,z) = \frac{v(z)}{t+(1-t)f(z)} . \tag{2.3}$$

Observe that $v(t,\cdot) \in C^{k,\alpha}$ (and not in $C^{k+1,\alpha}$). We then define the flow associated to this vector field as

$$\begin{cases} \frac{\partial}{\partial t}\Phi(t,x) = v(t,\Phi(t,x)) & t > 0 \\ \Phi(0,x) = x. \end{cases} \tag{2.4}$$

Note that because of the assumptions there exists a unique solution Φ: $[0,1] \times \bar{\Omega} \to \mathbb{R}^n$ in $\mathrm{Diff}^{k,\alpha}$ satisfying (2.4). We claim that $u(x) = \Phi(1,x)$ has all the desired properties, i.e., $u \in \mathrm{Diff}^{k,\alpha}(\bar{\Omega})$ and

$$\begin{aligned} \det\nabla u(x) &= f(x), & x \in \Omega \\ u(x) &= x & x \in \partial\Omega . \end{aligned} \tag{2.5}$$

(The aim of the next two steps will be to obtain $u \in \mathrm{Diff}^{k+1,\alpha}$ instead of $\mathrm{Diff}^{k,\alpha}$).

First observe that the fact that $u(x) = x$ on $\partial\Omega$ follows at once from the unicity of solutions of (2.4) and of the fact that $v(t,x) = 0$ if $x \in \partial\Omega$ (since $v = 0$ on $\partial\Omega$, c,f, (2.2)). It therefore remains to show that $\det \nabla u = f$ in Ω. This can be deduced from the fact that the following quantity

$$h(t,x) \equiv \det \nabla_x \Phi(t,x)[t + (1-t)f(\Phi(t,x))] \tag{2.6}$$

is independent of t (c.f. below for a proof). The result follows then from the fact that $h(1,x) = h(0,x)$. To prove that (2.6) is independent of t is obtained from (2.2) and the fact that the solution of (2.4) satisfies (as well known)

$$\frac{\partial}{\partial t}[\det \nabla_x \Phi(t,x)] = \det \nabla_x \Phi(t,x) \operatorname{div} v(t,\Phi(t,x)).$$

We refer for more details to Dacorogna-Moser [1].

Step 3: A classical fixed point argument allows to deduce from Theorem 2 the following: let $0 < \beta \leq \alpha < 1$, then there exists $\varepsilon = \varepsilon(\alpha,\beta,k,\Omega) > 0$ such that if $\|f-1\|_{C^{0,\beta}} \leq \varepsilon$ then there exists $u \in \mathrm{Diff}^{k+1,\alpha}(\bar{\Omega})$ satifying (2.5).

Step 4: We now are in a position to conclude the proof of the theorem. By density of C^∞ functions in $C^{k,\alpha}$ with the $C^{0,\beta}$ norm $0 < \beta < \alpha < 1$ (in general if $k = 0$ density does not occur in $C^{0,\alpha}$ norm) we can find $\tilde{f} \in C^\infty(\bar{\Omega})$, $\tilde{f} > 0$ on $\bar{\Omega}$ such that

$$\left\| \frac{f}{\tilde{f}} - 1 \right\|_{C^{0,\beta}} \leq \varepsilon \tag{2.7}$$

$$\int_\Omega \frac{f(x)}{\tilde{f}(x)}\, dx = \operatorname{meas} \Omega \tag{2.8}$$

where $\varepsilon > 0$ is as in Step 3.

We then define $b \in \text{Diff}^{k+1,\alpha}(\bar{\Omega})$ to be a solution (which exists by Step 3, (2.7) and (2.8)) of

$$\begin{cases} \det \nabla b(x) = \dfrac{f(x)}{\tilde{f}(x)} & \text{in } \Omega \\ b(x) = x & \text{on } \partial\Omega. \end{cases} \tag{2.9}$$

We then define $a \in \text{Diff}^{k+1,\alpha}(\bar{\Omega})$ to be a solution (which exists by Step 2, since $\tilde{f} \circ b^{-1} \in C^{k+1,\alpha}$) of

$$\begin{cases} \det \nabla a(y) = \tilde{f}(b^{-1}(y)) & y \in \Omega \\ a(y) = y & y \in \partial\Omega. \end{cases} \tag{2.10}$$

Note that the compatibility condition is satisfied since

$$\int_\Omega \tilde{f}(b^{-1}(y))dy = \int_\Omega \tilde{f}(x)\det \nabla b(x)dx = \int f(x)dx = \text{meas}\,\Omega.$$

Finally we get the desired diffeomorphism by setting $u = a \circ b$. □

Before sketching the proof of Theorem 2, we introduce some notations.

Notations:

Let $w \in C^1(\mathbb{R}^n; \mathbb{R}^{n(n-1)/2})$ with $w = (w_{ij})_{1 \leq i < j \leq n}$. We also let $w_{ij} = - w_{ji}$ if $i \geq j$. We then define

$$\text{curl}^* w = ((\text{curl}^* w)_j)_{1 \leq j \leq n} \in \mathbb{R}^n$$

by

$$(\text{curl}^* w)_j = \sum_{i=1}^{n} (-1)^{i+j} \frac{\partial w_{ij}}{\partial x_i} .$$

Remarks:

i) If $n = 2$ then $\text{curl}^* w = (\frac{\partial w}{\partial x_2}, - \frac{\partial w}{\partial x_1})$

and if $n = 3$ then $\mathrm{curl}^* w = \mathrm{curl}\, w$ (the usual curl).

ii) In terms of differential forms $(\mathrm{curl}^* w)_j$ are just the components of $d\alpha$ where α is an $(n-2)$ form with components w_{ij}.

iii) One always have

$$\mathrm{div}(\mathrm{curl}^* w) = 0 ,$$

which corresponds to $dd\alpha = 0$.

Proof of Theorem 2:

The idea for finding u satisfying

$$\begin{cases} \mathrm{div}\, u = f & \text{in } \Omega \\ u = 0 & \text{on } \partial\Omega \end{cases} \tag{2.10}$$

is to seek for solutions of the form

$$u = \mathrm{grad} v + \mathrm{curl}^* w. \tag{2.11}$$

We therefore solve (ν being the outward unit normal)

$$\begin{cases} \Delta v = f & \text{in } \Omega \\ \dfrac{\partial v}{\partial \nu} = 0 & \text{on } \partial\Omega \end{cases}$$

which, provided $\int f = 0$, has a unique $C^{k+2,\alpha}$ solution by the classical results. In view of (2.10) and (2.11) it therefore remains to find $w \in C^{k+2,\alpha}$ satisfying

$$\mathrm{curl}^* w = - \mathrm{grad}\, v \quad \text{on } \partial\Omega$$

knowing that $v \in C^{k+2,\alpha}$ and

$$\frac{\partial v}{\partial \nu} = 0.$$

Therefore Theorem 2 is reduced to finding $w \in C^{k+2,\alpha}$ satisfying

$$\text{curl}^* w = g \quad \text{on } \partial\Omega \tag{2.12}$$

knowing that $g \in C^{k+1,\alpha}$ with

$$<g;\nu> = 0 \quad \text{on } \partial\Omega \tag{2.13}$$

where $<\cdot;\cdot>$ stands for scalar product in $\mathbb{R}^n$.

The equations (2.12) can be further reduced to solving

$$\text{grad}\, w_{ij} = (-1)^{i+j}\,(g_j\nu_i - g_i\nu_j) \quad \text{on } \partial\Omega. \tag{2.14}$$

Indeed if the above holds on $\partial\Omega$ we have, recalling that $w_{ij} = -w_{ji}$,

$$(\text{curl}^* w)_j = \sum_{i=1}^{n} (-1)^{i+j} \frac{\partial w_{ij}}{\partial x_i} = \sum_{i=1}^{n} (g_j\nu_i - g_i\nu_j)\nu_i =$$

$$= g_j|\nu|^2 - <g;\nu>\nu_j = g_j$$

where we have used (2.13) as well as the fact that $|\nu| = 1$.

We therefore have reduced the problem to finding $w_{ij} \in C^{k+2,\alpha}$ satisfying

$$\text{grad}\, w_{ij} = g_{ij}\nu \quad \text{on } \partial\Omega \tag{2.15}$$

given $g_{ij} \in C^{k+1,\alpha}$ (here $g_{ij} = (-1)^{i+j}\,(g_j\nu_i - g_i\nu_j)$). We show how to solve (2.15) and find $w_{ij} \in C^{k+1,\alpha}$ (to obtain a more regular solution in $C^{k+2,\alpha}$, we refer to Dacorogna-Moser [1]). Observing that the gradient of the distance $d(x,\partial\Omega)$ is $-\nu$ whenever $x \in \partial\Omega$ we obtain a solution by setting

$$w_{ij} = -g_{ij}\,\phi(d(x,\partial\Omega))$$

where $\phi(0) = 0$, $\phi'(0) = 1$, $\phi \equiv 0$ outside a small neighbourhood of 0 and $\phi \in C^\infty$.

□

ACKNOWLEDGMENTS

I would like to thank M. Chipot and J. Saint Jean Paulin for organizing this nice conference in Metz.

REFERENCES

- S. Alpern [1]:
  New proofs that weak mixing is generic; Inventiones Math., Vol.32 (1976) 263-279.

- D.V. Anosov; A.B. Katok [1]:
  New examples in smooth ergodic theory, ergodic diffeomorphisms; Trans. Moscow Math. Soc., Vol.23 (1970) 1-35.

- J.M. Ball [1]:
  Convexity conditions and existence theorems in nonlinear elasticity; Arch. Rat. Mech. Anal., Vol.63 (1977) 337-403.

- A. Banyaga [1]:
  Formes volume sur les variétés à bord; Enseignement Math., Vol.20 (1974) 127-131.

- R. Coifman; P.L. Lions; Y. Meyer; S. Semmes [1]:
  Compacité par compensation et espaces de Hardy; C.R.A.S., Vol.309 (1989) 945-949.

- B. Dacorogna [1]:
  A relaxation theorem and its application to the equilibrium of gases; Arch. Rat. Mech. Anal., Vol.77 (1981) 359-385.

- B. Dacorogna [2]:
  "Direct methods in the calculus of variations", Springer, Berlin (1989).

- B. Dacorogna; J. Moser [1]:
  On a partial differential equation involving the Jacobian determinant; Ann. Inst. H. Poincaré, Anal. non-lin., Vol.7 (1990) 1-26.

- J. Moser [1]:
  On the volume element on a manifold; Trans. A.M.S. Vol.20 (1965) 286-294.

- S. Müller [1]:
  A surprising higher integrability property of mappings with positive determinant; Bull. A.M.S. Vol.21 (1989) 245-248.

- L. Tartar [1]:
  Private communication (1979).

- E. Zehnder [1]:
  Note on smoothing symplectic and volume preserving diffeomorphisms, Springer, Lect. Notes in Math., Vol.597 (1976) 828-855.

M J ESTEBAN

About the symmetry of positive solutions of nonlinear elliptic problems in symmetric domains

Université Pierre et Marie Curie
Laboratoire d'Analyse Numérique
75252 PARIS Cedex 05, FRANCE

ABSTRACT. This paper is a review of some results studying the symmetry properties of positive solutions of nonlinear elliptic problems in bounded or unbounded symmetric domains of $\mathbb{R}^N$. Different results are stated and some of the techniques used to prove them roughly described.

0. INTRODUCTION.

Let us consider the following problem

$$\begin{aligned} -\Delta u = f(u) \quad &\text{in} \quad \Omega, \\ au + b\frac{\partial u}{\partial n} = 0 \quad &\text{on} \quad \partial\Omega, \end{aligned} \tag{1}$$

where a and b are nonnegative real numbers, f is a locally lipschitzian real function such that $f(0) = 0$, n is the outward unit normal to Ω, Ω being any domain of $\mathbb{R}^N$.

The results obtained by Gidas, Ni and Nirenberg in [10] showed that for the Dirichlet problem, i. e. when $b = 0$, the nonnegative solutions of (1) reproduce the domain's symmetry when Ω is a ball, a rectangle, etc, and also when $\Omega = \mathbb{R}^N$ under adequate decay estimates for u at ∞.

These results suggest that there is a strong relationship between positivity and symmetry. Nevertheless negative results in this direction quickly showed that the situation is not so simple. In [4] nonradial positive solutions are proved to exist for certain annulus type domains and certain functions f, and in some cases in enormous amounts. In the case when Ω is bounded and $b = 0$, other positive and negative results in this direction can be found for instance in [3], [14], [7], [13]. [7] proves that for some annular domains only radial positive solutions exist. The case of the Neumann problem ($a = 0$) for the exterior of a ball is studied in [8,9]. In this case in which Ω is unbounded, positive nonradial solutions are again proved to exist.

The above seems to indicate that for some convex domains the domain's symmetry is respected by the positive solutions of the Dirichlet problem. But when Ω is no more convex all the possibilities are open and no general result seems to be true. However it will be very interesting to understand when the positive solutions of a general problem are symmetric. Of course, knowing that some solutions are symmetric allows to simplify many computations, proofs, numerical computations, etc, and in some cases it even enables one to obtain results which would be much more difficult to obtain for general solutions, with a priori no symmetry properties.

In this paper I intend to give a review of all these results and explain how they are proved and what kind of answers seem reasonable for the many remaining open problems.

The organisation of the paper is as follows. In Section I I deal with the cases in which the domain's symmetry is respected by the positive solutions, and I mainly explain which kind of results are proved in [10] and a rough description of the method of moving parallel planes, which is the proofs' main ingredient, is given.

In Section II I describe Coffman's result in [4] and its Li's extension ([14]) and also another result in [3] which uses a completely different proof to show, as Coffman or Li, that for some annulus type Ω's, there are nonradial positive solutions.

Section IV deals with the case of unbounded domains and I mainly describe the results in [8,9] in which for the exterior of a ball, radial and nonradial positive functions are proved to exist. In this case the domain's unboundedness creates an a priori lack of compactness that makes a lot of difference in the proof of the existence of nonradial positive solutions.

Finally in Section V a recent result of Dancer's ([7]) will lead us to discuss open problems and guess about their possible answers in view of the existing results.

I. When the positive solutions are necessarily symmetric.

In [10] Gidas, Ni and Nirenberg proved some results which have already been used in many different situations as well as the method employed to prove them. Various forms of the Maximum Principle plus a device of moving parallel planes to a critical position were at the base of these results. They gave not only new results about the symmetry properties of positive solutions of nonlinear elliptic and parabolic P.D.E.'s, but also a way to obtain new uniqueness results for this kind of problems.

The method of moving parallel planes is due to Alexandrov (see chapter 7 in [11]) and was again used by Serrin in [19]. It consists in considering a positive solution u of (1) and some hyperplane H which cuts Ω into two pieces; then via the Maximum Principle prove that u is somewhat monotone in the direction perpendicular to H in one of these pieces of Ω. Then move H in the perpendicular direction until it reaches some critical position and apply there again the Maximum Principle. This will give the desired information about u and about its derivative in the direction perpendicular to H.

The problem considered in [10] is the Dirichlet problem for (1), i. e. with $b = 0$.. It is also assumed that the nonlinearity f is of class C^1 and that the solutions u are in $C^2(\overline{\Omega})$. Then the following general result is proved

THEOREM I.1 ([10]). *Assume that Ω is a bounded domain of $\mathbb{R}^N$ which is symmetric about $x_1 = 0$ and convex in the x_1 direction. Then if $u \in C^2(\overline{\Omega})$ is a positive solution of (1) with $b = 0$ and if $f \in C^1(\overline{\Omega})$, u is symmetric with respect to x_1 and $\frac{\partial u}{\partial x_1} < 0$ for $x_1 > 0$.*

A corollary of this theorem is the best known result in [10]; it states the following

THEOREM I.2 ([10]). *Let Ω be the ball : $\{x \in \mathbb{R}^N \ ; \ r = |x| < R\}$ and assume that u and f satisfy the same assumptions as in the previous result. Then u is radially symmetric and*

$$\frac{\partial u}{\partial r} < 0 \quad , \qquad \text{for } 0 < r < R .$$

REMARK I.3. These results are very strong because almost no assumption is made on f. On the other hand, they still hold when the laplacian is replaced by some more general elliptic operators and the nonlinearity f may also depend on $x \in \Omega$ (see [10] for the precise assumptions that are needed in these cases). Moreover an alternative proof of this result is given in [16]. This alternative proof applies only to the particular case $N = 2$, $f \geq 0$, but on the other hand it allows weaker assumptions on f and u. The method used in [16] relies on the Schwarz symmetrization, the use of the classical isoperimetric inequality (see for instance [1]) and the well-known Pokhozaev's identity ([18]).

As explained above, in the case of annular domains there are examples of non radially symmetric positive solutions of (1). When the techniques used to prove Theorems I.1-2 are applied to this case the outcome is the following

THEOREM I.4 ([10]). *Let $u > 0$ be a C^2 solution of (1) with Dirichlet boundary conditions ($b = 0$) in a ring shaped domain*

$$R' < r = |x| < R .$$

Then

$$\frac{\partial u}{\partial r} < 0 \quad \text{for} \quad \frac{R+R'}{2} \leq |x| < R .$$

The above result will not be of course contradictory with the examples given in Sections II-III, in which the existence of nonradial positive solutions is proved for ring shaped domains. Nevertheless it gives information about the possible location of critical points of positive solutions of (1) in an annular domain in the case of Dirichlet boundary conditions.

The above results have been extended in [2] to the case of sector-like domains and mixed boundary conditions. One of the problems considered in [2] is the following

Let $(r, \theta_1, \cdots, \theta_{N-1})$ be the polar coordinates in $\mathbb{R}^N$, $r \geq 0$, $\theta_i \in [0, \pi]$, $i = 1, \cdots, N-2$, and $\theta_{N-1} \in [0, 2\pi)$. Then define the spherical sector $\Sigma(\alpha, R)$ of radius R and amplitude $\alpha \in (0, 2\pi)$ by

$$\Sigma(\alpha, R) = \{x \in \mathbb{R}^N \; ; \; 0 < |x| = r < R, \; \theta_i \in (0, \pi) \text{ and } \theta_{N-1} \in (0, \alpha)\} .$$

The boundary of this sector consists of two pieces, $\Gamma_0 = \{x \in \partial\Sigma(\alpha, R) \; ; \; |x| = R\}$ and $\Gamma_0 = \{x \in \partial\Sigma(\alpha, R) \; ; \; \theta_{N-1} = 0 \text{ or } \theta_{N-1} = \alpha\}$.

Then the P.D.E. problem considered is

$$-\Delta u = f(u) \quad \text{in} \quad \Sigma(\alpha, R)$$

together with the boundary conditions

$$u = 0 \quad \text{on} \quad \Gamma_0 \quad , \quad \frac{\partial u}{\partial n} = 0 \quad \text{on} \quad \Gamma_1 .$$

and for this problem the symmetry result is the following

THEOREM I.5 ([2]). *Let $0 < \alpha \leq \pi$ and $f \in C^1(\mathbb{R}_+)$. If $u \in C^2(\overline{\Sigma}(\alpha, R))$ is a solution of the above boundary value problem and if u is positive in $\Sigma(\alpha, R)$, then u is spherically symmetric and*

$$\frac{\partial u}{\partial r} < 0 \quad \text{for} \quad 0 < r < R .$$

This theorem (and others in [2]) are proved by using again several Maximum Principles and moving parallel planes in an astucious way. The above theorem gives another example of respect of the domain's symmetry by the positive solutions of a nonlinear elliptic problem. But together with this result one finds in [2] an example showing that Theorem I.5 does not hold any more as soon as the sector is larger than half a ball, that is, when $\pi < \alpha < 2\pi$. Hence once again we see that this kind of results are not so general, and that making general conjectures is not an easy task.

II. A first (direct) method to find non symmetric positive solutions.

In [4] Coffman considered the following case:

$$N = 2\,, \quad b = 0\,, \quad f(s) = |s|^{2q}s,\ q \text{ being a positive integer,} \tag{2}$$

$$\Omega = \Omega(r, c) = \{(x, y) \in \mathbb{R}^2\,;\ r^2 < x^2 + y^2 < r^2 + c^2\}. \tag{3}$$

For this particular case Coffman proved the following result, which in particular implies the existence of nonradial positive solutions of (1) when the product rc is large enough.

THEOREM II.1 ([4]). *Let $c_0, \mu > 0$ and the natural numbers q (≥ 1) and M be given. There exists an r_1 such that if $r \geq r_1$ and $c_0 \leq c \leq \mu r$ then problem (1), together with assumptions (2) and (3), has at least M rotationally non-equivalent positive solutions.*

Coffman used a variational principle to prove this theorem. He associated with problem (1)-(3) the *Raileigh quotient*

$$J(u) = \frac{\left(\int_\Omega \left(|\nabla u|^2 + u^2\right)\, dx\, dy\right)^{q+1}}{\int_\Omega u^{2q+2} dx\, dy}$$

$J(u)$ is defined for all non-zero functions in $H_0^1(\Omega)$. Then for all positive integer k, $V_k(r, c)$ is defined as the positive cone in the subspace of $H_0^1(\Omega)$ that consists of all the functions u which are invariant under rotations of $2\pi/k$ around the origin; in the same way $V_\infty(r, c)$ is defined as the positive cone of the subspace of $H_0^1(\Omega)$ which consists of all radially symmetric functions.

For k a positive integer or $k = \infty$ let

$$j_k = j_k(r,c) = \inf\{J(v) \, : \, v \in V_k(r,c) \setminus \{0\}\}.$$

That j_k is achieved for every integer k and for $k = +\infty$ follows in particular from the fact that the V_k's are convex closed cones in $H_0^1(\Omega)$. Then upper estimates are obtained for j_k and lower estimates for j_∞. Finally an estimate for the k's such that $j_k < j_\infty - 1$ and a number theoretical argument complete the proof of Theorem II.1. This is of course a very quick description of how this theorem is proved, but gives an idea of the methods used in [4].

Let us point out that Coffman's proof can be considerably simplified by a result obtained by Kawohl in [12] which shows that the minimizer of J in V_k does not belong to V_m for any proper multiple m of k.

The above result applies to dimension $N = 2$ and particular nonlinearities, $f(s) = |s|^{2q}s$, q being a positive integer . But at the end of the paper Coffman remarks that Theorem II.1 can be extended in a straigthforward way to to higher even dimensions and also to the case $f(s) = |s|^{p-1}s$, p being a positive real number less than $N/(N-2)$.

In [14] Li extends Coffman's result. Basically he proves that Theorem II.1 still holds for $f(s) = |s|^{p-1}s$ or $f(s) = |s|^{p-1}s - s$ and either $N = 2$ and $p > 1$ or $N \geq 3$ and $1 < p < \frac{N+2}{N-2}$. More general nonlinearities f are also considered, as sums of subcritical powers ($p < \frac{N+2}{N-2}$) or supercritical powers. But for more general nonlinearities no general result seems to be known.

The method used by Li is a refined form of Coffman's, but the basic ideas are the same.

III. An indirect method to find non symmetric positive solutions.

In [3] the existence of symmetric and non symmetric positive solutions for problem (1) arises from a discussion about uniqueness of positive solutions of this problem. The case of an annulus Ω is again considered together with the assumption: $f(s) = |s|^{p-1}s$, $1 < p$. The proof in [3] goes as follows. The positive numbers S_p and Σ_p are defined by

$$S_p = \inf\{||\nabla u||_2^2 \; ; \; u \in H_0^1(\Omega), \, ||u||_{p+1} = 1\},$$

$$\Sigma_p = \inf\{||\nabla u||_2^2 \; ; \; u \in H_0^1(\Omega), \, u \text{ radial} \, , \, ||u||_{p+1} = 1\},$$

where by $||\cdot||_q$ we denote the $L^q(\Omega)$-norm.

It is easy to check that the functions $p \to S_p$ and $p \to \Sigma_p$ are continuous. Moreover since Ω is an annulus, for every $p > 1$ the embedding

$$\{u \in H_0^1(\Omega)\,;\; u \text{ radial }\} \hookrightarrow L^{p+1}(\Omega)$$

is compact for every $p > 1$. Hence it is trivial to find a solution of Σ_p. Moreover after some stretching and by using the Maximum Principle, a positive solution of (1), u_p, can be found. The same argument shows that S_p is also achieved (by a positive funcion) for $p < \frac{N+2}{N-2}$.

On the other hand

$$S_{\frac{N+2}{N-2}} < \Sigma_{\frac{N+2}{N-2}}. \tag{4}$$

Indeed $S_p \leq \Sigma_p$ for all p and equality cann't hold in (4), since otherwise $S_{\frac{N+2}{N-2}}$ would be achieved, which is impossible (see for instance [18] and [3]). Therefore for p less than $\frac{N+2}{N-2}$, but close to it, $S_p < \Sigma_p$, which proves the existence of both symmetric and nonsymmetric positive solutions of (1) in this case.

As we will see in Section V below, this result cann't be extended much further, since as is proved in [7], for every $1 < p < \frac{N+2}{N-2}$ there are annulus of external radius 1 and small internal radius such that the positive solution of (1) is unique (and then necessarily radially symmetric).

IV. Unbounded domains.

The existence of at least one solution of (1) for Ω bounded in $\mathbb{R}^N$ and f a nonlinear function with a subcritical growth at infinity (less quicky than $|s|^{\frac{N+2}{N-2}}$) is easy to obtain by using variational arguments and the compactness of the embedding $H_0^1(\Omega) \hookrightarrow L^q$ for $q < \frac{N+2}{N-2}$. As we already saw in Section III this is not the case when $q = \frac{N+2}{N-2}$, since then the above embedding is no more compact. The same happens when Ω is unbounded. In this section the results of [8,9], where problem (1) in exterior domains and with Neumann boundary conditions is considered, will be presented. This problem can be written as follows

$$\begin{aligned} -\Delta u + u &= |u|^{p-1}u \quad \text{in} \quad \mathbb{R}^N \backslash \overline{B_R}\,, \\ \frac{\partial u}{\partial n} &= 0 \quad \text{on} \quad \partial B_R\,, \end{aligned} \tag{5}$$

where $1 < p < \frac{N+2}{N-2}$ and $R > 0$. A priori and due to the lack of compactness the existence of solutions of (5) is not immediate. For instance if instead of considering (5) we consider the same P.D.E. equation as in (5), but with a Dirichlet boundary condition: $u = 0$ on ∂B_R, then we may look at a problem similar to (6) below, but replacing $u \in H^1$ by $u \in H_0^1$. Then this new variational problem does not have any solution. This can be easily proved by an argument involving an unique continuation principle.

To solve (5) we define the following variational problem

$$I_R = \inf_{u \in H^1(\mathbb{R}^N \setminus \overline{B}_R)} \left\{ \int_{\mathbb{R}^N \setminus \overline{B}_R} \left(|\nabla u|^2 + |u|^2\right) dx \, ; \int_{\mathbb{R}^N \setminus \overline{B}_R} |u|^{p+1} \, dx = 1 \right\}. \tag{6}$$

Of course, solutions of (6) are solutions of (5) multiplied by some constant, since (5) is the Euler-Lagrange equation associated with (6). On the other hand and since $f(0) = 0$, it is easy to see that we can choose a nonnegative solution of (6) whenever this problem is solvable. Finally, once a nonnegative solution of (5) exists, by the Maximum Principle it will be positive in Ω. I could also look for other critical points of the energy in (6) instead of looking only for minimizers, but as we will see below, considering problem (6) will be enough for our purposes of finding symmetric and nonsymmetric solutions of (5).

In order to state the existence results proved in [8,9], let us define

$$I = \inf_{u \in H^1(\mathbb{R}^N)} \left\{ \int_{\mathbb{R}^N} \left(|\nabla u|^2 + |u|^2\right) dx \, ; \int_{\mathbb{R}^N} |u|^{p+1} \, dx = 1 \right\}. \tag{7}$$

The question of the existence of solutions of problem (6) is then answered by the following theorems.

THEOREM IV.1 ([8,9]). *If for some R the following strict inequality holds*

$$I_R < I \tag{8}$$

then I_R is achieved.

The large inequality in (8) is easy to prove (see for instance [15]). That the strict inequality is sufficient to insure that I_R is achieved is typical in this kind of problems. Actually what the strict inequality implies is the relative compactness in $H^1(\Omega)$ of all minimizing sequences for (6). This result is proved by the concentration-compactness method (see [15]).

THEOREM IV.2 ([8,9]). *For all* $R > 0$ *we have* $I_R < I$.

This theorem is proved by considering a solution of (7) and using it as a test function to estimate I_R.

The solutions proved to exist in the above two theorems are positive, but nothing is known about their uniqueness or their symmetry properties. Let us call them u_R. Symmetric (radial) positive solutions of (5), w_R, also exist for every $R > 0$. Indeed if we define the new variational problem

$$J_R = \inf_{u \in H^1_*(\mathbb{R}^N \setminus \overline{B}_R)} \left\{ \int_{\mathbb{R}^N \setminus \overline{B}_R} (|\nabla u|^2 + |u|^2)\, dx \; ; \int_{\mathbb{R}^N \setminus \overline{B}_R} |u|^{p+1}\, dx = 1 \right\}, \qquad (9)$$

where $H^1_*(\mathbb{R}^N \setminus \overline{B}_R) = \{u \in H^1(\mathbb{R}^N \setminus \overline{B}_R)\,; u(x) = \overline{u}(|x|),\ x \in \mathbb{R}^N\}$, then J_R is readily solvable. In effect, the embedding $H^1_*(\mathbb{R}^N \setminus \overline{B}_R) \hookrightarrow L^{p+1}$ is now compact. Therefore standard variational techniques yield a solution of (9), w_R.

The question is now whether u_R and w_R are the same function or not. Moreover if they are different functions, what about the respective energies I_R and J_R?

These questions are answered in [8,9] and the answer goes in the same direction of Coffman's result (see Theorem II.1 above). Some precise estimates are necessary to prove the following result

PROPOSITION IV.3 ([8,9]). *The function* $R \to J_R$ *is continuous and monotone increasing for all* $R > 0$*, the value at* 0 *being* $J_0 = I$.

The monotonicity of J_R is obtained by considering a solution w_R and by using it to build up a good "test" function for the problem $J_{R'}$ with R' less than R but close to it. A precise knowledge of w_R's behavior at ∂B_R is then necessary to prove the above proposition.

COROLLARY IV.4 . *For every* $R > 0$ *there are at least a radial and a nonradial positive solution of (5).*

REMARK. Other results about the behaviour of J_R and I_R are available in [8,9], but I will not state them here, since they are irrelevant for this paper's purposes.

REMARK. In some sense, and with the notation of Section II, for a fixed R, $J_R = j_\infty$ and $I_R = \inf\{j_k \; ; \; k \text{ integer} \; , \; k \geq 1\}$. Coffman's result showed that for a large family of (bounded) annulus, there are some (sometimes many) k's for which $j_k < j_\infty$. Theorems IV.1-2 prove the same inequality. Coffman's result was even more accurate, since it yielded the existence of multiple distinct solutions, corresponding to different k's. On the other hand, problem (5) is more difficult to deal with since the unboudedness of Ω creates additional problems as far as the existence of (multiple) solutions is concerned.

V. Other results and open problems.

After having looked at all the previous results it appears clearly that the relationship between positivity and symmetry is not easy to understand. Gidas, Ni and Nirenberg's results pointed in one direction, the respect of the domain's symmetry by the positive solutions of (1). But almost all the results proved since seem to give more and more examples of desobeyance to this rule. However one may observe that all the negative results have been proved for annulus or exterior domains. Hence in nonconvex domains. On the other hand, the results described in Section I, which are positive as far as the symmetry of positive solutions is concerned, are also defined in convex domains. Moreover let us point out that the positive solutions in annulus or exteriors of balls which are found in Sections II-IV are not radial, but still have some symmetry property: they are axially symmetric with respect to some radial line going from the origin to some point in the annulus' boundary..

Of course in the case of convex domains one can easily understand that some kind of "symmetrization" leaving the domain invariant is possible. And a "good symmetrization" decreases energy. Hence if a solution exists and it is obtained from minimization, another one, symmetric, can be obtained by symmetrizing the problem. This is all very vague but gives a hint about which kind of problems are better adapted to have only positive solutions which have their symmetry properties.

A recent result by Dancer (see its announcement in [7]) shows that when the domain does not seem "well adapted" for the positive solutions to be symmetric, it still may have only positive solutions which respect its symmetry. He considers problem (1) with $b = 0$ and $f(s) = |s|^{p-1}s$, i.e.

$$\begin{aligned} -\Delta u &= |u|^{p-1}u \quad \text{in } \Omega, \\ u &= 0 \quad \text{on } \partial\Omega, \end{aligned} \tag{10}$$

where Ω is a bounded annular domain in $\mathbb{R}^N$, $N \geq 2$ and $1 < p < \frac{N+2}{N-2}$. A corollary of the main result in [7] states the following

THEOREM V.1 ([7]). *Let $\Omega = \{x \in \mathbb{R}^N;\ \varepsilon < |x| < 1\}$. If ε is small, (10) has a unique positive solution (necessarily radially symmetric).*

For radially symmetric solutions and with no restriction on ε, this (uniqueness) result was already proved in [17]. But of course in the case of an annulus the restriction on ε is necessary, since otherwise there would be a contradiction with the result of [3] described in Section III.

The above result follows from the fact (see [5,6,7]) that if good a priori estimates and some nondegeneracy property are available, then the number of positive solutions is continuous with respect to continuous changes in the domain's shape.

In [7] can also be found the complementary result to Theorem V.1, which states a result similar to that in [3] (see Section III):

THEOREM V.2 ([7]). *Let Ω be as in Theorem V.1. There exists ε_0 depending on p such that, if $\varepsilon_0 < \varepsilon < 1$, then (10) has a radially symmetric solution and at least one nonradially symmetric solution.*

This theorem is proved, as above, by considering the corresponding variational problems and comparing the infima's (or maxima) values for symmetric or nonsymmetric functions.

Let us conclude this discussion by saying that all the above results, in one or the other direction, seem to indicate that it will be difficult to understant fully the problem of respect of the domain's symmetry by the positive solutions of nonlinear elliptic partial differential equations. And as was already said in the Introduction, understanding this problem would be useful for knowing which kind of symmetry is available for positive solutions, which could considerably simplify the presentation of many results, their proofs and some numerical computations.

REFERENCES

[1] C. BANDLE. *Isoperimetric inequalities.* Pitman, London (1980).

[2] H. BERESTYCKI, F. PACELLA. Symmetry properties for positive solutions of elliptic equations with mixed boundary conditions. Preprint.

[3] H. BREZIS, L. NIRENBERG. Positive solutions of nonlinear elliptic equations involving critical Sobolev exponents. Comm. Pure Appl. Math. **36** (1983), p. 437-477.

[4] C.V. COFFMAN. A non-linear boundary value problem with many positive solutions. J. Diff. Eq. **54**, (1984), p. 429-437.

[5] E.N. DANCER. The effect of domain shape on the number of positive solutions of certain nonlinear equations, J. Diff. Eq. **74**, (1988), p. 120-156.

[6] E.N. DANCER. On the influence of domain shape on the existence of the large solutions of some superlinear problems. Math. Ann., to appear.

[7] E.N. DANCER. Certain weakly nonlinear partial differential equations on annuli. Nonlinear Anal. TMA, **14(5)**, (1990), p. 441-442.

[8] M.J. ESTEBAN. Rupture de symétrie pour des problèmes de Neumann surlinéaires dans des ouverts extérieurs. Note C.R.A.S. **308**, Série I, (1989), p. 281-286.

[9] M.J. ESTEBAN. Nonsymmetric ground states of symmetric variational problems. To appear in Comm. Pure Appl. Math.

[10] B. GIDAS, W.M. NI, L. NIRENBERG. Symmetry of positive solutions of nonlinear elliptic equations in $\mathbb{R}^N$. Advances in Math. Studies, **7A**, (1981), p. 369-402.

[11] H. HOPF. *Lectures on differential geometry in the large.* Stanford University, (1956).

[12] B. KAWOHL. On the isoperimetric nature of a rearrangement inequality and its consequences for some variational problems. LCDS Report 84-4, Providence, (1984).

[13] B. KAWOHL, M. LONGINETTI. On radial symmetry and uniqueness of positive solutions of a degenerate elliptic eigenvalue problem. Z. angew. Math. Mech., **68(5)**, (1988), p. 459-460.

[14] Y.Y. LI. On second order nonlinear elliptic equations. Dissertation, New York Univ., (1988).

[15] P.L. LIONS. The concentration-compactness method in the Calculus of variations. The locally compact case. Part I: Ann. IHP. Anal. Non Lin. **1** (1984), p. 109-145.

[16] P.L. LIONS. Two geometrical properties of solutions of semilinear problems. Appl. Anal., **12**, (1981), p. 267-272.

[17] W.M. NI, R. NUSSBAUM. Uniqueness and nonuniqueness for positive radial solutions of $\Delta u + f(u,r) = 0$. Comm. Pure Appl. Math., **38**, (1985), p. 67-108.

[18] S. POKHOZAEV. Eigenfunctions of the equation $\Delta u + \lambda f(u) = 0$, Math. Dokl., **6**, (1965), p. 1408-1411.

[19] J. SERRIN. A symmetry problem in potential theory. Arch. Rat. Mech., **43**, (1971), p. 304-318.

M GRÜTER

Geometric measure theory and minimal surfaces

I. Let us start by considering a typical problem in the calculus of variations.

Suppose $\Gamma \subset \mathbb{R}^3$ is a simple closed Jordan curve. Find a surface $M \subset \mathbb{R}^3$ which has Γ as its boundary, $\partial M = \Gamma$, such that its area $|M|$ is as small as possible.

This problem is known as Plateau's Problem, and a solution surface is called a minimal surface. In other words, minimal surfaces are extrema of a variational problem, that of minimizing area.

A necessary condition for something to be an extremum is the vanishing of the first variation. Hence -in the case of minimizing area- a solution has to satisfy a certain partial differential equation of second order,the associated Euler-Lagrange- equation. For a smooth solution u of this equation this corresponds to the fact that the image of u -regarded as a surface (submanifold) in the surrounding space- has mean curvature zero. On the other hand, it can easily be shown that any surface of vanishing mean curvature is locally (in space) area minimizing.

The Plateau Problem mentioned above was solved around 1930 independently by Douglas and Radó in the classical parametric context. They proved the existence of a mapping $x\colon D \to \mathbb{R}^3$, D being the unit disk in $\mathbb{R}^2$, such that x parametrizes a disk-type minimal surface of least area spanning Γ. The solution x is constructed in such a way that it minimizes Dirichlet's integral

$$D(x) := \iint_D |x_u|^2 + |x_v|^2 \, du \, dv$$

as well, that is x is harmonic: $\Delta x^i = 0$, i = 1, 2, 3. Thus, x is in fact real analytic in the interior. This settles the question of analytic regularity.

However, the question of geometric regularity has to be dealt with separately. Is the solution embedded or at least immersed, does it have true of false branch points? Note, that sometimes there are no embedded disk-type

surfaces spanning Γ (for example for knotted curves). In these cases the Douglas-Radó solution must necessarily have self-intersections.
For details concerning the parametric approach the reader is referred to the encyclopedic monograph by Nitsche [NJCC] and the recent book by Struwe [SM].

II. Let us now turn to Geometric Measure Theory (GMT).
To motivate the different notions of surfaces that have been introduced in GMT consider the following example.
Suppose $U \subset \mathbb{R}^{n+k}$ is open, $M = M^n \subset U$ is an n-dimensional oriented C^1-submanifold which is assumed to have locally finite measure, i.e. $\mathcal{H}^n(M \cap K) < \infty$ for any compact subset K of $\mathbb{R}^{n+k}$.
Here, $\mathcal{H}^n$ denotes n-dimensional Hausdorff measure in $\mathbb{R}^{n+k}$ which coincides with the area-element of M.
To such a submanifold M we can associate a certain linear functional $[\![M]\!]$ acting on $\mathcal{D}^n(U)$:

$$[\![M]\!](\omega) := \int_M \omega \ , \quad \omega \in \mathcal{D}^n(U).$$

Here,
$\mathcal{D}^n(U) = \{\omega\colon \omega \text{ a } C^\infty\text{-differential-n-form on } U, \text{ spt } \omega \text{ a compact subset of } U\}$,
and integration is understood in the usual sense of differential forms on oriented manifolds.
In this way M can be thought of as a continuous linear functional on $\mathcal{D}^n(U)$.
One now defines the space of n-*currents* on U

$$\mathcal{D}_n(U) := [\mathcal{D}^n(U)]^*$$

as the dual space -with respect to the topology of uniform convergence of all derivatives on compact subsets- of $\mathcal{D}^n(U)$.

The *boundary* ∂T of such a $T \in \mathcal{D}_n(U)$ is then defined by

$$\partial T(\omega) := T(d\omega) \text{ for } \omega \in \mathcal{D}^{n-1}(U), \ n \geq 1.$$

Here, d denotes exterior differentation.For n=0 one defines $\partial T=0$.
With these definitions one obviously has $\partial T \in \mathcal{D}_{n-1}(U)$ as well as $\partial^2 T=0$.
Note, that $\mathcal{D}_0(U)$ is the space of distributions on K introduced by Laurent Schwartz. One might think of n-currents as vector-valued distributions.
If M as above is such that ∂M satisfies the same conditions it is an easy consequence of Stokes' Theorem that

$$\partial\, [\![M]\!] = [\![\partial M]\!].$$

To generalize the notion of area one introduces the notion of *mass*, that is for an open set $W \subset U$ and $T \in \mathcal{D}_n(U)$ one defines

$$\mathbf{M}_W(T) := \sup \{T(\omega) : |\omega| \le 1, \ \text{spt}\ \omega \subset W\}.$$

One checks that for $W \subset\subset U$ (i.e. $\overline{W}$ compact $\subset U$) and M as above

$$\mathbf{M}_W([\![M]\!]) = \mathcal{H}^n(M \cap W).$$

Of course, with these definitions the space of n-currents enjoys very good compactness properties with respect to mass-bounded families (Banach-Alaoglu). For geometric purposes however the notion of a general n-current is much too weak.

A subspace that has turned out to be very useful in the geometric calculus of variations is the space of *rectifiable n-currents*. An n-current $T \in \mathcal{D}_n(U)$ is said to be *rectifiable*, if

$$T(\omega) = \int \langle \omega(x), \zeta(x) \rangle \, \theta(x) \, d\mathcal{H}^n(x), \quad \omega \in \mathcal{D}^n(U).$$

Here, $M \subset U$ is a countably n-rectifiable set, i.e. a subset of a countable union of n-dimensional C^1-submanifolds together with a set of vanishing $\mathcal{H}^n$-measure. It can be shown that for such an M there exists an approximate tangent space T_xM for $\mathcal{H}^n$-a.e. $x \in M$.

For these $x \in M$ the function $\zeta(x)$ is an $\mathcal{H}^n$-measurable choice of an orientation for T_xM, and $\theta(x) \ge 0$ is also supposed to be $\mathcal{H}^n$-measurable.

One writes $\mu_T = \mathcal{H}^n \llcorner \theta$ and $T = \tau(M, \theta, \zeta)$.

An important special class inside the rectifiable n-currents is the space of *integer multiplicity rectifiable n-currents.* They are characterized by the fact that $\theta_T(x) \in \mathbb{N} \cup \{0\}$ for $\mathcal{H}^n$-a.e. $x \in M$. These currents have a certain geometric structure and for them one has the fundamental

EXISTENCE THEOREM (Federer and Fleming [FF], 1960)

Suppose, that $\{T_j\}$ is a sequence of integer multiplicity rectifiable n-currents in U. If

$$\sup_j \{\mathbf{M}_W(T_j) + \mathbf{M}_W(\partial T_j)\} < +\infty$$

for every open $W \subset\subset U$, then there exists a subsequence $\{T_{j'}\}$ and an integer multiplicity rectifiable n-current T in U such that

$$T_{j'} \to T \quad \text{as } j' \to \infty.$$

Furthermore, if each T_j is mass-minimizing then T is mass-minimizing as well and $\mu_{T_{j'}} \to \mu_T$ in the sense of Radon-measures.

Remark:

The existence of a general n-current T as the limit of some subsequence is of course trivial. The main point in the Existence Theorem above (also called Compactness Theorem, Closure Theorem) is the assertion, that T is again integer multiplicity. Thus, in some sense Federer's and Fleming's Theorem can be considered as a first regularity result.

Concerning the regularity theory for area-minimizing integer multiplicity rectifiable n-currents let me only state the results (without being very precise) in the codimension one case.

REGULARITY THEOREM (k=1)

Suppose that T is area minimizing and denote by sing T the set of interior singularities (away from spt ∂T).

Then

$$\mathcal{H}\text{-dim (sing T)} \le n-7.$$

Moreover, if ∂T is smooth, then there are no singularities near or on spt ∂T.

Remark:

The first part of the theorem is the final result of the work of a number of mathematicians during the 60's. Here, let me only mention the work of Fleming, De Giorgi, Almgren, Simons, Bombieri-De Giorgi-Giusti, and Federer, see the bibliography in [SL]. The surprising result concerning the boundary regularity was obtained by Hardt and Simon [HS] and appeared in 1979.

For the Plateau Problem in $\mathbb{R}^3$ (mentioned in section I) these results imply the following

THEOREM
Suppose $\Gamma \subset \mathbb{R}^3$ is a smooth, simple, closed, rectifiable Jordan arc. Then there exists an embedded smooth oriented submanifold $M \subset \mathbb{R}^3$ with $\partial M = \Gamma$ such that M has least area among all oriented surfaces S with $\partial S = \Gamma$.

Finally let me mention that analogous regularity results for free boundary problems have been obtained in [GM 1,2,3].
For further information on GMT the reader is referred to [FH], the standard monograph, as well as [MF], [HS2], and [SL].

III. There is another class of surfaces in GMT that have turned out to be useful, the so called varifolds that were introduced in the 60's by Almgren.
A general n-*varifold* in $U \subset \mathbb{R}^{n+k}$ is a Radon measure on $U \times G(n+k,n)$.
Here, the more geometrical objects corresponding to integer multiplicity currents are the *integer multiplicity (rectifiable)* n-*varifolds*.
Suppose, $T = t(M,\theta,\zeta)$ is given, then $V = v(M,\theta)$ is the corresponding varifold, and for $f \in C_c^0\ (U \times G(n+k,n),\ \mathbb{R})$ we have

$$V(f) = \int_M f(x,T_xM)\ \theta(x)d\mathcal{H}^n(x).$$

The following result, obtained by Allard in 1972, is fundamental.

THEOREM ([AW1], abstract interior regularity)
Suppose, that V is a stationary rectifiable n-varifold with $\theta_V \geq 1$ μ_V-a.e. .
Furthermore, assume that $x \in \operatorname{spt} \mu_V$, $\rho > 0$, $\delta > 0$ such that

$$\mu_V(B_\rho(x)) \leq 2(1-\delta)\omega_n\rho^n = 2(1-\delta)\ \mathcal{H}^n(B_\rho^n).$$

There exists $\varepsilon = \varepsilon(n,k,\delta) > 0$ such that if the *excess*

$$E(x,\rho,\mathbb{R}^n) := \frac{1}{2}\,\rho^{-n} \int_{B_\rho(x)} ||P_{T_xM} - P_{\mathbb{R}^n}||^2 d\mu_V$$

satisfies $E(x,\rho,\mathbb{R}^n) < \varepsilon$ then

$$B_{\rho/2}(x) \cap \text{spt } V = \text{graph } u$$

for some $u\colon \mathbb{R}^n \cap B_{\rho/2}(P_{\mathbb{R}^n} x) \to \mathbb{R}^k$; the $C^{1,\alpha}$-norm of u can be estimated in terms of the excess.

Remarks:

(1) The fact that V is assumed to be stationary means that the first variation of V -which can be defined for varifolds- is zero, and for smooth surfaces corresponds to the vanishing of the mean curvature.

(2) Here, we have chosen to state a rather special case of Allard's Theorem in a somewhat technical version that will be used in the last section of this paper. For a detailed discussion and proof the reader is referred to [SL].
An analogous result for a fixed boundary was also obtained by Allard [AW2].
In 1986 Jost and the author presented a version of Allard's Theorem for the case of free boundaries [GJ1]. This was used by the same authors to show the existence of an embedded minimal disk M inside a convex body $K \subset \mathbb{R}^3$ such that M is perpendicular to ∂K [GJ2]. In 1989 Jost [JJ] succeeded in proving the existence of three geometrically different minimal disks.

The most famous application of GMT-techniques in minimal surface theory is probably the proof of the Positiv Mass Conjecture in general relativity given by Schoen and Yau in 1979 [SY]. In 1989 Schoen [SR] published a proof of a generalization of the Positive Mass Conjecture, The Positive Energy Theorem (PET), in dimensions $n \le 7$. In [KE1] Kuwert carried out a detailed proof of the PET along the lines given in [SR].
A crucial step in these proofs is the construction of a stable embedded minimal hypersurface. However, the method of construction given in [KE1] can also be used to handle other problems for minimal surfaces. This is done in [KE2] for example for Exterior Plateau Problems as in [TY].

IV. In this final section let me describe in some detail an application of the method used in [KE1] to an

EXTERIOR FREE BOUNDARY PROBLEM

Suppose $M \subset \mathbb{R}^3$ is a compact, smooth surface without boundary, and let $\nu \in S^2$

be a unit vector.

Find a minimal surface $\Sigma \subset \mathbb{R}^3$ such that the following is true:

(a) $\partial\Sigma \subset M$;

(b) Σ intersects M orthogonally along $\partial\Sigma$;

(c) there is a plane $\Pi \perp \nu$, a disk D in Π, and a function u such that Σ = graph u over $\Pi \sim D$;

(d) Σ has the asymptotic normal ν at infinity.

For the construction of Σ we proceed as follows.

Without loss of generality we may assume $\nu = (0,0,1)$ and that M and $\mathbb{R}^2 = \{(\xi,0)\colon \xi=(\xi_1,\xi_2),\ \xi_i \in \mathbb{R}\}$ intersect transversely.

Let $p\colon \mathbb{R}^3 \to \mathbb{R}^2$ be the orthogonal projection $:\xi = p(x)$, and for $\sigma > 0$ denote by $D_\sigma = \{\xi \in \mathbb{R}^2\colon |\xi| < \sigma\}$ the open disk of radius σ centered at the origin in $\mathbb{R}^2$.

Choose $R_0 > 0$ such that $M \subset B_{R_0}(0)$.

For $\sigma > R_0$ let $\Gamma_\sigma := [\![\partial D_\sigma]\!] \in \mathcal{D}_1(\mathbb{R}^3)$ and consider the class

$$\mathcal{C}(M, \sigma) := \{T \in \mathcal{D}_2(\mathbb{R}^3)\colon T \text{ integer multiplicity, } \operatorname{spt} T \text{ compact, } \operatorname{spt}(\partial T - \Gamma_\sigma) \subset M\}.$$

We want to minimize mass in the class $\mathcal{C}(M,\sigma)$.

This is possible by various compactness theorems, in particular [FH;5.1.6] and [GM1; Satz 1.8].

Thus, we get $\Sigma_\sigma \in \mathcal{C}(M,\sigma)$ such that

$$\mathbf{M}(\Sigma_\sigma) = \inf \{\mathbf{M}(T)\colon T \in \mathcal{C}(M,\sigma)\} .$$

Since M and $\mathbb{R}^2$ intersect transversely we obviously have

$$(1) \qquad \mathbf{M}(\Sigma_\sigma) \le \pi\sigma^2 - a_0$$

for some a_0 not depending on σ.

Next, note that since the 2-current Σ_σ - $[\![D_\sigma]\!]$ has no boundary in the cylinder over the domain $\mathbb{R}^2 \sim \overline{D_{R_0}}$ the same is true for the projection $p_\#(\Sigma_\sigma - [\![D_\sigma]\!])$. But $\mathbb{R}^2 \sim \overline{D_{r_0}}$ is connected, and hence the constancy theorem applies to give

$$p_\#(\Sigma_\sigma - [\![D_\sigma]\!]) = c\ [\![\mathbb{R}^2 \sim \overline{D_{R_0}}]\!]$$

for some integer c.

The fact that

$$\operatorname{spt}\ (\Sigma_\sigma - [\![D_\sigma]\!]) \subset \overline{B_\sigma(0)}$$

implies $c = 0$.

From that we conclude

$$p^{-1}\{\xi\} \cap \operatorname{spt}\ \Sigma_\sigma \neq \emptyset$$

for any $\xi \in \mathbb{R}^2$ such that $R_0 < |\xi| < \sigma$.

As a consequence we deduce for any measurable set

$$U' \subset \{\xi \in \mathbb{R}^2 \colon R_0 < |\xi| < \sigma\} =: A(R_0,\sigma)$$

that

(2) $$\mathcal{H}^2(\operatorname{spt}\ \Sigma_\sigma \cap p^{-1}(U')) \geq \mathcal{L}^2(U').$$

For $U \subset A(R_0, \sigma)$, $U' = A(R_0, \sigma) \sim U$, and the weight measure $\mu_\sigma = \mathcal{H}^2 \llcorner \theta_\sigma$ we use (1) and (2) to derive the following estimate:

$$\pi\sigma^2 - a_0 \geq \mathbf{M}(\Sigma_\sigma) =$$

$$= \mu_\sigma(p^{-1}(U)) + \mu_\sigma(p^{-1}(U')) + \mu_\sigma(p^{-1}(\overline{D_{R_0}})) \geq$$

$$\geq \mu_\sigma(p^{-1}(U)) + \mathcal{L}^2(U') + \mu_\sigma(p^{-1}(\overline{D_{R_0}})).$$

This can be written as

$$\mu_\sigma(p^{-1}(U)) + \mu_\sigma(p^{-1}(\overline{D_{R_0}})) \le$$

(3)

$$\le \pi\sigma^2 - \mathcal{L}^2(A(R_0,\sigma) \sim U) - a_0 = \pi R_0^2 + \mathcal{L}^2(U) - a_0 = \mathcal{L}^2(U) + a_1.$$

Here, $a_1 = a_1(a_0,R_0) := \pi R_0^2 - a_0$.

Applying this inequality for $U = \overline{D_\tau} \cap A(R_0,\sigma)$ and $R_0 < \tau < \sigma$ we deduce

$$\mathbf{M}(\Sigma_\sigma \llcorner p^{-1}(\bar{D}_\tau)) = \mu_\sigma(\{|\xi| \le \tau\}) \le$$

(3′)

$$\le \pi\tau^2 - \pi R_0^2 + a_1 = \pi\tau^2 - a_0 .$$

Note, that this estimate is independent of σ.

To apply the compactness theorem of Federer and Fleming we still have to estimate the mass of the free boundary.

This can be done by combining a result from [GM2] with the estimate (3′).

More precisely, from [GM2;Corollary 3.2] we get a constant C depending only on M such that

$$\mathbf{M}(\partial\Sigma_\sigma - \Gamma_\sigma) \le C\ [1 + \mathbf{M}(\Sigma_\sigma \llcorner p^{-1}(\overline{D_{R_0+1}}))]$$

(4)

$$\le C\ [1 + \pi(R_0+1)^2 - a_0]$$

$$= C'(M,\ R_0,\ a_0)\ .$$

The compactness theorem now yields the existence of a subsequence $\{\Sigma_{\sigma_j}\}$, $\lim_{j\to\infty} \sigma_j = \infty$, and of an integer multiplicity 2-current Σ in $\mathbb{R}^3$ such that

$$\Sigma_{\sigma_j} \to \Sigma \qquad \text{as } j \to \infty.$$

Moreover, since the Σ_{σ_j} are minimizing, their limit Σ has the following minimizing property.

For any ball B in $\mathbb{R}^3$ and any $X \in \mathcal{D}_2(B)$ such that X is integer multiplicity, spt X ⊂ B, spt ∂X ⊂ M, we have

$$\mathbf{M}_B(\Sigma) \le \mathbf{M}_B(\Sigma + X).$$

Additionally $\mu_{\Sigma_{\sigma_j}} \to \mu_\Sigma$ as $j \to \infty$ in the sense of Radon measures.

Note, that by construction spt(∂Σ) ⊂ M.

Since the weight measures converge the estimate (3) implies

(5) $$\mathbf{M}(\Sigma \llcorner p^{-1}(\bar{D}_\tau)) \le \pi\tau^2 - a_0$$

for any $\tau > R_0$.

Next, we want to apply Allard's regularity theorem for large balls to improve the convergence of the Σ_σ's to Σ .

First, observe that Σ_σ is a smooth connected minimal surface with boundary. In fact, the smoothness follows from the corresponding codimension one regularity results mentioned above (interior regularity, regularity near the fixed boundary ∂D_σ, regularity at the free boundary. The connectedness of Σ_σ follows from the fact that otherwise the component of Σ_σ not containing ∂D_σ could be subtracted which would decrease area.

Note, that the regularity theory also implies $\theta_\sigma \equiv 1$.

Since $M \subset B_R(0)$ and $\partial\Sigma_\sigma \subset (M \cup \partial D_\sigma)$ the height of Σ_σ over $\mathbb{R}^2$ can be controlled by R_0:

(6) $$|z| \le R_0 \text{ for } x = (\xi,z) \in \operatorname{spt} \Sigma_\sigma .$$

Consider now any point $x = (\xi,z) \in \operatorname{spt} \Sigma_\sigma$ such that

$$|\xi| = \tau \ge R_0 + \rho \quad , \ \rho + \tau \le \sigma.$$

It is well known that the excess (of Σ_σ over $\mathbb{R}^2$) can be estimated by the height as follows

$$E(x, \rho/2) \le \frac{c}{\rho^2} \int_{B_\rho(x)} \left[\frac{\operatorname{dist}(y-x,\mathbb{R}^2)}{\rho}\right]^2 d\mu_\sigma(y) \quad .$$

Using (6) we can deduce

$$E(x,\rho/2) \le \frac{c}{\rho^2} \ \frac{R_0^2}{\rho^2} \ \mu_\sigma(B_\rho(x)).$$

For $U = D_\rho(p(x)) = D_\rho(\xi)$ we apply (3) and get

$$\mu_\sigma(B_\rho(x)) \le \mu_\sigma(p^{-1}(U)) \le \mathcal{L}^2(U) + a_1 = \tag{7}$$
$$= \pi\rho^2 + a_1 = \left(1 + \frac{a_1}{\pi\rho^2}\right)\pi\rho^2 .$$

Thus, the excess is controlled by

$$E(x,\rho/2) \le \frac{1}{\rho^2}\, c\, R_0^2\left(\pi + \frac{a_1}{\rho^2}\right) . \tag{8}$$

Now, choose $\rho_0 = \rho_0(c,R_0,a_0) > 0$ so large that for $\rho \ge \rho_0$ we have

$$a_1 < \frac{1}{2}(\pi\rho^2) \quad \text{in (7)}$$

and so that the right hand side of (8) is less than the $\varepsilon > 0$ in Allard's regularity theorem.

Thus, all the assumptions in Allard's theorem have been verified and we conclude

$$\text{spt } \Sigma_\sigma \cap \{R_1 < |\xi| < \tfrac{2}{3}\sigma\} = \text{graph } u_\sigma$$

for $R_1 = R_0 + \rho_0$ and a suitable $C^{1,\alpha}$-function

$$u_\sigma\colon A(R_1, \tfrac{2}{3}\sigma) \to \mathbb{R}.$$

Here, we have again used the height bound (6) for spt Σ_σ (w.l.o.g. $\rho_0 > 2 R_0$). Furthermore, the $C^{1,\alpha}$-norm of u_σ can be estimated by the excess, in fact, since u_σ solves the minimal surface equation, we conclude that u_σ is real analytic.

The representation of Σ_σ (outside some compact set) as the graph of a solution of the minimal surface equation together with good a-priori-estimates for the gradient can be used in a standard way to improve the convergence of the Σ_σ's to Σ .

Thus, we have shown that Σ is the graph of a real analytic function u (outside some compact set), and the gradient estimates for u_σ imply that $\nabla u(\xi)$ tends uniformly to zero as $|\xi| \to \infty$, i.e. Σ is asymptotic to a plane parallel to $\mathbb{R}^2$. Of course the same proof works for dimensions $n \le 6$. These arguments are carried out in detail in [KE1], [KE2].

Let us finally mention that Σ itself cannot be a plane since this would contradict the estimate (5). We have therefore solved this exterior free boundary problem and the solution is non-trivial.

BIBLIOGRAPHY

[AW1] W.K. Allard: On the first variation of a varifold.
Ann.Math. **95**(1972),417-491.

[AW2] W.K. Allard: On the first variation of a varifold: boundary behavior.
Ann.Math.**101**(1975),418-446.

[FF] H. Federer, W.H.Fleming:Normal and integral currents.
Ann.Math.**72**(1960),458-520.

[FH] H. Federer: Geometric measure theory.
Springer (1969).

[GJ1] M. Grüter, J. Jost: Allard type regularity results for varifolds with free boundaries.
Ann. Sc. Norm. Sup. Pisa (4)**13**(1986), 129-169.

[GJ2] M. Grüter, J. Jost: On embedded minimal disks in convex bodies.
Anal.non lin.**3**(1986),345-390.

[GM1] M. Grüter: Regularität von minimierenden Strömen bei einer freien Randbedingung.
Habilitationsschrift,Düsseldorf(1985).

[GM2] M. Grüter: Optimal regularity for codimension one minimal surfaces with a free boundary.
Man.math. **58**(1987),295-343.

[GM3] M. Grüter: Regularity results for minimizing currents with a free boundary.
J. Reine Angew. Math. **375/376** (1987), 307-325.

[HS1] R. Hardt, L. Simon: Boundary regularity and embedded solutions for the oriented Plateau problem.
Ann.Math. **110**(1979),439-486.

[HS2] R. Hardt, L. Simon: Seminar on geometric measure theory.
DMV Seminar, Band 7;Birkhäuser (1986).

[JJ] J. Jost: Embedded minimal surfaces in manifolds diffeomorphic to the three-dimensional ball or sphere.
J.Differential Geometry **30**(1989),555-577.

[KE1] E.C. Kuwert: Der Minimalflächenbeweis des Positive Energy Theorem.
Vorlesungsreihe Nr. 14, SFB 256, Bonn (1990).

[KE2] E.C. Kuwert: Eingebettete Lösungen von Außenraumproblemen für Minimalflächen.
Thesis, Bonn (1990).

[MF] F. Morgan: Geometric measure theory - A beginner's guide.
Academic Press (1988).

[NJCC] J.C.C. Nitsche: Vorlesungen über Minimalflächen.
Springer (1975).
Lectures on minimal surfaces, Vol.1.
Cambridge University Press (1989).

[SL] L. Simon: Lectures on geometric measure theory.
Proc. CMA **3**, Austr. Nat.Univ. Canberra (1983).

[SM] M. Struwe: Plateau's problem and the calculus of variations.
Math.Notes **35**, Princeton University Press (1988).

[SR] R. Schoen: Variational theory for the total scalar curvature functional for Riemannian metrics and related topics.

In: M.Giaquinta (Ed.), Topics in Calculus of Variations. Lecture Notes in Math.1365, Springer (1989).

[SY] R. Schoen, S.T. Yau: On the proof of the Positive Mass Conjecture in general relativity. Comm.Math.Phys. **65**(1979),45-76.

[TY] F. Tomi, R. Ye:The exterior Plateau problem. Preprint,Universität Heidelberg (1988).

Michael Grüter
Fachbereich Mathematik
Universität des Saarlandes
D-6600 Saarbrücken

B KAWOHL

From p-Laplace to mean curvature operator and related questions

The motivation for my lecture comes from an optimal design problem in the theory of torsion. Let $\Omega \in \mathbb{R}^2$, a simply connected domain, be the prescribed cross-section of a cylindrical bar. Two linearly elastic materials with different shear moduli should fill out Ω in such a way, that the torsional rigidity of the bar is maximized. This problem has a fairly obvious solution if Ω is a circular disk, then the weak material should be located on a circular disk D inside Ω, while the stronger material should be placed on the annular outer ring. If one perturbs Ω only slightly, say from a disk to an ellipse with small eccentricity, D can no longer be a set with C^1 boundary, see [GKR,KSW]. In fact one cannot expect a classical solution of this problem to exist. There are weak formulations, however, which lead in a natural way to a certain free boundary problem. In this new free boundary problem there are subdomains on which a function u satisfies

$$\operatorname{div}(\nabla u) = const. \tag{0.1}$$

and other subdomains, on which it satisfies

$$\operatorname{div}(|\nabla u|^{-1}\nabla u) = const. \tag{0.2}$$

Equations (0.1)(0.2) are the Euler equations to a (relaxed) variational problem, namely

$$\min_{v\in W_0^{1,2}(\Omega)} J(v) = \int_\Omega \{j(|\nabla v|) - v\}dx, \tag{0.3}$$

where $j(s)$ is piecewise linear or quadratic and is plotted in Figure 1.

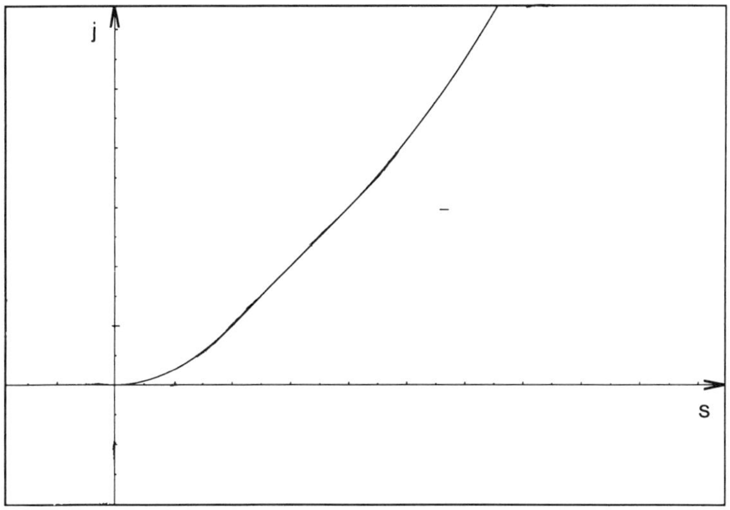

Figure 1: $j(s)$ for the optimal design problem

This should be compared with the corresponding functions $j(s) = \frac{1}{2}\mu s^2$ for a linear elastic material, or

$$j(s) = \begin{cases} s^2 & \text{if } s \leq 1, \\ +\infty & \text{if } s > 1, \end{cases}$$

in the elastic-plastic torsion problem, see [T].

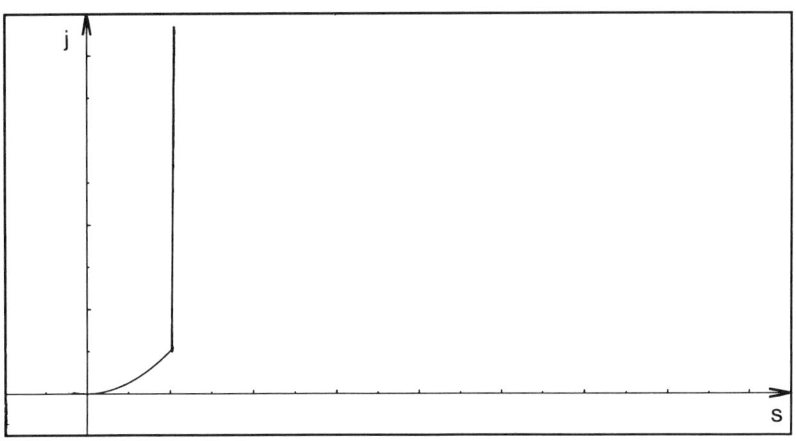

Figure 2: $j(s)$ for the elastic plastic torsion problem

To get some understanding for the operator in (0.2) I studied a model problem which connects equations (0.1) and (0.2) through a parameter p. In this model problem $j(s) = \frac{1}{p}s^p$.

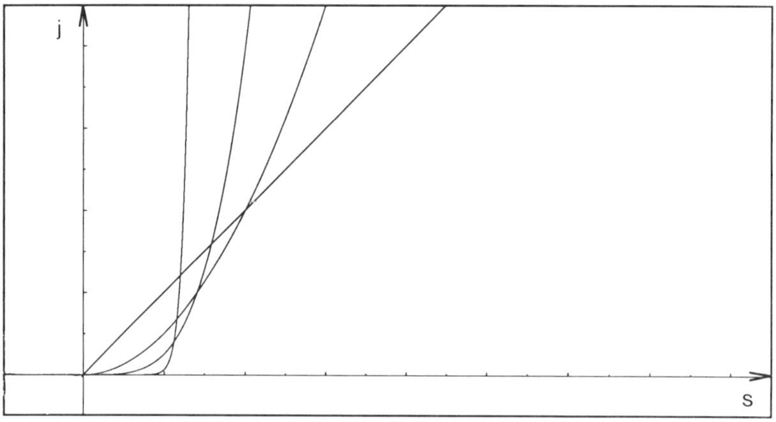

Figure 3: $j(s)$ for the model problem

As it turns out, the limit $p \to 1$ has some surprises. And the limit $p \to \infty$ is equally interesting. In addition to the results which I presented in Metz in December 1988, there are two (apparently new) results in these notes, namely Theorems 4 and 5. It is my pleasure to thank M.Chipot and J. Saint Jean Paulin for organizing such fine meetings. My research was financially supported in part by the DFG.

§1. The limit $p \to \infty$ in $\Delta_p u_p = -1$

For fixed $p \in (1, \infty)$ consider the problem

$$-\Delta_p u_p = 1 \qquad \text{in } \Omega, \tag{1.1}$$

$$u_p = 0 \qquad \text{on } \partial\Omega. \tag{1.2}$$

Here $\Omega \subset I\!R^N$ is a bounded domain, p is a parameter which varies between 1 and ∞ and

$$\Delta_p u = \text{div}(|\nabla u|^{p-2}\nabla u). \tag{1.3}$$

A function $u \in W_0^{1,p}(\Omega)$ is called a solution to (1.1) (1.2) if and only if

$$\int_\Omega |\nabla u|^{p-2} \nabla u \nabla \phi dx = \int_\Omega \phi dx \qquad \text{for any } \phi \in C_0^\infty(\Omega).$$

Problem (1.1) (1.2) models torsional creep. If a cylindrical bar of constant cross section $\Omega \in I\!R^2$ is subjected to torsion, the function u is called warping function, $|\nabla u|$ represents stress, and the shear strain energy is given by $|\nabla u|^{p-1}$, see [L,PP,P] for details.

For $1 < p < \infty$ the existence of a unique solution u_p to (1.1) (1.2) is well known. In general $u \in C^{1,\alpha}$ for some $\alpha \in (0,1)$, but unless $p = 2$, u_p need not be of class C^2 in points where $\nabla u_p = 0$, see the radial solutions in §2. If $|\nabla u(x_0)| > 0$, however, then $u \in C^\infty$ in a neighbourhood of x_0 in Ω [T].

For convex Ω it was recently shown that $u_p^{(p-1)/p}$ is a concave function [S]. This suggests that $u_\infty = \lim_{p\to\infty} u_p$, if it exists, should be concave. As we shall see $u_\infty = d(x, \partial\Omega)$ confirms the conjectured concavity.

The limiting case $p \to \infty$ is of special interest in applications, because it models perfectly plastic torsion, while the case $p = 2$, in which Δ_p is the usual Laplace operator, represents perfectly elastic torsion. There are interesting results on elastic-plastic torsion problems [Ti] which suggest that

$$\lim_{p\to\infty} u_p(x) = d(x\partial\Omega), \tag{1.4}$$

the distance function to $\partial\Omega$. Another result which supports (1.4) is due to Payne and Phillippin [PP, p.454] and states that

$$\lim_{p\to\infty} \int_\Omega u_p(x) dx = \int_\Omega d(x, \partial\Omega) dx. \tag{1.5}$$

Notice that (1.5) is weaker than (1.4). Theorem 1 (below) states that (1.4) holds uniformly in Ω. In fact the convergence takes place in C^α with $\alpha < 1$. Under the additional assumption that Ω is pseudoconvex, i.e. that the boundary $\partial\Omega$ is smooth and has nonnegative mean curvature $H(x)$ the convergence takes even place in $C^{0,1}(\Omega)$, see Theorem 2.

In the sequel it will be useful to recall that (1.1) represents the Euler equation associated with the functional

$$J_p(v) = \int_\Omega \{\frac{1}{p}|\nabla v|^p - v\}dx. \tag{1.6}$$

For $1 < p < \infty$ this functional is well-defined and strictly convex on $W_0^{1,p}(\Omega)$.

It is furthermore instructive to inspect an identity for the p-Laplace operator Δ_p which holds under the assumption that $|\nabla v(x)| > 0$. In those points the level surface $\{x \in \Omega \,|\, v(x) = t\} = \partial\Omega_t$ has a well-defined exterior normal $n(x) = -\frac{\nabla v(x)}{|\nabla v(x)|}$ and mean curvature $H(x)$. If we denote differentiation with respect to $n(x)$ by the subscript $_n$, we can easily calculate, using curvilinear coordinates

$$\begin{aligned}\Delta_p(v) &= |\nabla v|^{p-2}(\Delta v + (p-2)v_{nn}) \\ &= |\nabla v|^{p-2}((p-1)v_{nn} + (N-1)H(x)v_n).\end{aligned} \tag{1.7}$$

For $p = 2$ this is the well-known representation of the Laplace operator (see e.g. [Ta,K2])

$$\Delta v = v_{nn} + (N-1)H(x)v_n.$$

This formula is not only of aesthetic value. It gives heuristic ideas as to the behaviour of u_p for $p \to \infty$. If one looks at (1.7) and (1.1) for $p \to \infty$, one expects for $\hat{u} = \lim_{p\to\infty} u_p$ that

$$\hat{u}_{nn} = 0, \tag{1.8}$$

because $p - 1 \to \infty$. This heuristic expectation is in fact justified, because the function $\hat{u} = d(x, \partial\Omega)$ satisfies (1.8) a.e. in the sense of distributions.

Theorem 1. *Let $\Omega \subset I\!R^n$ be a bounded domain and let $u_p \epsilon W_0^{1,p}(\Omega)$ be the solution to* (1.1)(1.2) *for $1 < p < \infty$. Then*

$$\lim_{p\to\infty} u_p(x) = d(x, \partial\Omega) \qquad \textit{uniformly in } \Omega. \tag{1.9}$$

For the proof one shows that the family $\{u_p\}_{p\to\infty}$ is uniformly bounded in $C^\alpha(\Omega)$ for any $\alpha \in (0,1)$. Under geometric assumptions on Ω one can even prove stronger results.

Theorem 2. *In addition to the assumptions of Theorem 1, let Ω be strictly pseudoconvex, with smooth boundary, i.e. $\partial\Omega$ is of class C^2 and has positive mean curvature $H_M \geq H(x) \geq h_m > 0$. Then*

$$|\nabla u_p(x)| \leq [\frac{1}{(N-1)H_m}]^{1/(p-1)} \qquad \text{in } \Omega, \tag{1.10}$$

i.e. the family $\{u_p\}_{p>2}$ is uniformly Lipschitz continuous and (1.9) *holds.*

To prove Theorems 1 and 2 one has to apply variational arguments, a priori estimates and suitable maximum principles. I just want to sketch the main idea of the proof of Theorem 1, the detailed prooofs can be found in [K4]. It is apparent from (1.5) and (1.6) that $\lim_{p\to\infty} J_p(u_p) \leq \lim_{p\to\infty} J_p(d(x,\partial\Omega))$ exists, and that

$$\lim_{p\to\infty} \int_\Omega \{\frac{1}{p}|\nabla v|^p\}dx \tag{1.11}$$

is finite. If $u_\infty = \lim_{p\to\infty} u_p$ exists, then (1.11) implies $|\nabla u_\infty| \leq 1$ a.e. in Ω. So u_∞ has to stay below $d(x,\partial\Omega)$ (in the weak inequality sense). This and (1.5) implies $u_\infty = d(x,\partial\Omega)$. So one has to show that the limit $u_\infty = \lim_{p\to\infty} u_p$ exists, and this is done by some standard estimates and compactness theorems.

Incidentally, in the meantime (July 1990) I have been informed that T. Bhattacharya, E.DiBenedetto and J.Manfredi (Limits as $p \to \infty$ of $\Delta_p u_p = f$ and related extremal problems, manuscript) have obtained results similar to Theorem 1.

§1. The limit $p \to 1$ in $\Delta_p u_p = -1$

Again we consider Problem (1.1) (1.2); but this time we send $p \to 1$. If we write down (1.6) for $p = 1$ and try to minimize it over $W_0^{1,1}(\Omega)$, there are problems with the existence of a solution in $W_0^{1,1}(\Omega)$. This space is not reflexive. A standard way to overcome this difficulty is to work in the space $BV(\Omega)$, the functions of bounded variation, see [G,AT]. In that case J_1 can be extended to $BV(\Omega)$ and

$$J_1(v) = \int_0^\infty \{P(\Omega_t, I\!R^N) - A(\Omega_t)\}dt \tag{2.1}$$

for $0 \le v \in BV(\Omega)$. Notice that the boundary condition $v = 0$ is hidden in the fact that I wrote $P(\Omega_t, I\!R^N)$ and not $P(\Omega_t, \Omega)$ in (2.1). Here $\Omega_t = \{x \in \Omega \mid v(x) > t\}$ is a level set of v and $P(\Omega_t, B)$ is the perimeter of Ω_t in B. Furthermore, $A(\Omega_t)$ denotes "area" or "volume", i.e. the N-dimensional Lebesgue measure of Ω_t. One can rewrite (2.1) as

$$J_1(v) = \int_\Omega |Dv|dx - \int_\Omega vdx + \int_{\partial\Omega} vd\mathcal{H}_{N-1} \tag{2.2}$$

For the ease of writing let me introduce the notation

$$P(t) := P(\Omega_t, I\!R^N), \qquad A(t) := A(\Omega_t)$$

Now we are in a position to explain why the case $p \to 1$ is of independent interest. The isoperimetric inequality [B,D] states that

$$P(t) \ge c(N)\, A(t)^{1-\frac{1}{N}}, \tag{2.3}$$

with

$$c(N) = N\pi^{1/2}2^{1/2}[N\Gamma(\frac{N}{2})]^{-1/N} \tag{2.4}$$

so that

$$A(t) < [c(N)]^N \qquad \Longrightarrow \qquad P(t) - A(t) > 0. \tag{2.5}$$

If Ω and thus $A(0)(\ge A(t))$ is sufficiently small, this implies that $u_1 \equiv 0$ is the only minimizer of (1.7). It furthermore suggests that

$$\lim_{p\to 1} u_p(x) = 0 \qquad \text{in } \Omega. \tag{2.6}$$

I will state (2.6) under a smallness assumption on Ω in Theorem 3 below. One can also show (see [K4]) that the smallness assumption on $A(0)$ is sufficient but far from necessary for (2.6). If, however, Ω is too large, then $J_1(v)$ can be unbounded below. One can easily see this from the radial case. Let Ω be a disk in $I\!R^2$ of radius $R > 2$, and let χ_Ω be its characteristic function. Then $J_1(\chi_\Omega) = 2\pi R - \pi R^2 < 0$ and $J_1(\lambda\chi_\Omega) = \lambda J_1(\chi_\Omega)$. This suggests that for such a domain

$$\lim_{p\to 1} u_p(x) = +\infty \qquad \text{in } \Omega. \tag{2.7}$$

One can prove (2.7) for balls in $I\!R^N$ of radius $R \geq N$. In fact, in [K4] the solutions of (1.1)(1.2) on balls of radius R are explicitly calculated. They are given by

$$u(r) = u(0)\,(1 - (\frac{r}{R})^q) \quad \text{with } q = \frac{p}{p-1} \tag{2.8}$$

and

$$u(0) = (1/N)^{1/(p-1)}\,\frac{p-1}{p}\,R^{p/(p-1)} = \frac{N}{q}(\frac{R}{N})^q$$

Notice that $q \to \infty$ if $p \to 1$ (and $q \to 1$ if $p \to \infty$). Let us rewrite (2.8) more explicitly as

$$u_p(r) = N\frac{1}{q}[(\frac{R}{N})^q - (\frac{r}{q})^q] := u(r,q) \tag{2.9}$$

Now one sees that (2.6) holds for $R < N$ while (2.7) holds for $R \geq N$. So one cannot expect u_p to have monotone dependence on p, even though there is some continuous dependence (see [Li]). By the way, for $p \to \infty$, (2.9) confirms (and has in fact served to guess) Theorem 1.

These simple considerations suggest that there is no "nice" limit $\tilde{u} = \lim\limits_{p\to 1} u_p$. Another one is a formal inspection of (1.7) for $p = 1$, which yields $(N-1)H\tilde{u}_n = \tilde{u}_n$ or

$$H(x) \equiv \frac{1}{N-1},$$

i.e. all level surfaces of $\tilde{u}$ should have the same constant mean curvature. This requirement anticipates the result that $\lim\limits_{p\to 1} |\nabla u_p| = 0$ a.e. in Ω, and that u_p develops transition layers as $p \to 1$.

It is obvious from (2.9) and from a comparison argument, that (2.6) holds if $\Omega \subset I\!R^N$ is contained in a ball of radius N. A stronger result is given below. To understand its formulation and proof it is necessary to define the Schwarz-symmetrizations Ω^* and u^* of a domain Ω and a function u. Ω^* is a ball in $I\!R^N$ with center in the origin and same N-dimensional Lebesgue measure as Ω. For a given continuous function u, u^* is the function which satisfies $\{x \in \Omega \mid |u(x)| > t\}^* = \{x \in \Omega^* \mid u^*(x) > t\}$ for every $t \geq 0$. More details on these definitions can be found e.g. in [B,K1].

Theorem 3. *Suppose that* $\Omega \subset I\!R^N$ *has the property*

$$\Omega^* \subset B_R(0) \quad \text{with} \quad R \leq N. \tag{2.10}$$

Then the estimate

$$||u_p||_{L^\infty(\Omega)} \leq N\frac{p-1}{p}(\frac{R}{N})^{p/(p-1)} \tag{2.11}$$

holds. In particular (2.6) *holds.*

To prove this theorem we compare the solutions of (1.1)(1.2) with solutions U_p of

$$-\Delta_p U_p = 1 \qquad \text{in } \Omega^*,$$

$$U_p = 0 \qquad \text{on } \partial\Omega^*.$$

A theorem of Giarusso [Gi] says that

$$||u_p||_{L^\infty(\Omega)} \leq ||U_p||_{L^\infty(\Omega^*)}.$$

Since U_p is explicitly known from (2.9), one can infer (2.11) and (2.6). ∎

It should be noted that assumption (2.10) is only sufficient but far from necessary for the validity of (2.6). Other sufficient criteria on Ω which cover domains that violate (2.10) are provided in [K4]. What these results show is that for general domains we cannot tell whether the family $\{u_p\}_{p>1}$ converges or diverges as $p \to 1$. However, the following can be stated.

Theorem 4. *Let $\Omega \subset I\!R^n$ be a bounded domain and let $u_p \epsilon W_0^{1,p}(\Omega)$ be the solution to (1.1)(1.2) for $1 < p < \infty$. Then at least one of the families $\{u_p\}_{p>1}$ or $\{v_p\}_{p>1}$, where*

$$v_p(x) = \frac{u_p(x)}{||up||_{L^\infty(\Omega)}},$$

has a (sub)sequence which is bounded in $BV(\Omega)$ as $p \to 1$ and which converges in $L^1(\Omega)$ and pointwise a.e. in Ω.

For the proof we distinguish two cases. If $\overline{\lim\limits_{p\to 1}} \int u_p(x)dx$ is finite, then

$$||u_p||_{BV(\Omega)} \leq ||\nabla u||_{L^p(\Omega)}\ |\Omega|^{1/q}, \qquad \left(\frac{1}{p} + \frac{1}{q}\right) = 1$$

is uniformly bounded due to the Euler equation. Thus $\{u_p\}_{p>1}$ is bounded in $BV(\Omega)$ as $p \to 1$. And otherwise

$$\begin{aligned} ||v_p||_{BV(\Omega)} &\leq ||u_p||^{-1}_{L^\infty(\Omega)}\ ||\nabla u||_{L^p(\Omega)}\ |\Omega|^{1/q} \\ &= ||u_p||^{-1}_{L^\infty(\Omega)} \left(\int u_p(x)\, dx\right)^{1/p} |\Omega|^{1/q}. \end{aligned} \tag{2.12}$$

But after taking the $\overline{\lim\limits_{p\to 1}}$ in (2.12) the theorem follows, since by assumption

$$\left(\int u_p(x)\, dx\right)^{1/p} \leq \int u_p(x)\, dx$$

on a subsequence. Therefore the right hand side of (2.12) can be estimated, using the uniform boundedness of v_p. ∎

The natural question arises if there are domains Ω such that $u_p(x) \to \infty$ a subset D of Ω while $u_p(x)$ remains finite on the complement of D, as $p \to 1$. Can one say anything about such subsets D? A partial answer is provided by the following result:

Theorem 5. *Suppose that* $\overline{\lim\limits_{p\to 1}} \int u_p(x)dx$ *is infinite and* Ω *is convex. Then for every sequence* v_p *which converges according to Theorem 4, the set of blow up points, i.e. points* $x \in \Omega$ *in which* $\overline{\lim\limits_{p\to 1}}\, u_p(x) = +\infty$ *is convex (modulo a nullset).*

For the proof we consider the sequence v_p defined in Theorem 4 and recall that $u_p^{(p-1)/p}$ is a concave function [S]. In particular $u_p(x)$ is log concave for any $1 < p < \infty$, i.e. $u_p(tx + (1-t)y) \geq u_p(x)^t u_p(y)^{1-t}$ for any $x, y \in \Omega$ and any $t \in [0,1]$. Consequently v_p is logconcave. Since logconcavity is preserved under pointwise convergence, Theorem 5 follows from Theorem 4. ∎

It has become apparent in the above, in particular in the discussion of (2.1) that the following geometric problem plays a crucial role in the analysis of the limiting case $p \to 1$.

Problem P1. Given $\Omega \subset I\!R^N$ find a subset $D \subset \bar{\Omega}$ such that $P(D) - A(D) \leq P(E) - A(E)$ for any measurable $E \subset \bar{\Omega}$.

Here as in the sequel $P(D)$ is the perimeter of D in $I\!R^N$ and $A(D)$ the N-dimensional Volume of D. Problem P1 has been treated for special domains Ω by J.Steiner as early as 1842, see [St].

One can think of the boundary of D as a free boundary inside a "container" Ω. Obstacle problems of this nature occur in microgravity physics, when one studies a liquid with surface tension in a partly empty tank Ω. In that case the volume A of the region which is occupied by the liquid is given and its surface P has to be minimized. Or one can imagine a rubber balloon which is partly inflated inside a container Ω and which has frictionless contact with the boundary. Yet other related variational problems from a model for phase transitions are described in [M1,M2]. Finally, Problem P1 shows up when one tries to describe plate failure under pressure, see [Ke].

Related to Problem P1 is the following obstacle problem.

Problem P2. Given $\Omega \subset I\!R^N$ find a solution $u \in BV(\Omega)$ such that

$$J_1(u) \leq J_1(v)$$

for any nonnegative $v \in BV(\Omega)$ with $||v||_{L^\infty(\Omega)} \leq 1$.

In (2.6) and (2.7) we have seen that Problem P2 can have the trivial solution $u \equiv 0$ or the solution $u(x) = \chi_\Omega(x)$, depending on the structure and size of Ω. This was due to the fact that in Problem P1 the solution D was the empty set and hence $P(D) - A(D) = 0$ or $D = \Omega$ and $P(D) - A(D) < 0$. It will be demonstrated below that Problem P2 can have solutions of the type $u(x) = \chi_D$ with $\emptyset \neq D \neq \Omega$. For the case that Ω is a large square in $I\!R^2$, this is shown in detail after the proof of Theorem 13.

Theorem 6. *Problems P1 and P2 are equivalent in the following sense:*

i) If $u \in BV(\Omega)$ is a solution to Problem P2, then almost every level set of u is a solution to Problem P1.

ii) If D is a solution to Problem P1, then $u = \chi_D$ is a solution to Problem P2.

Proof: We recall the notation of $P(t)$ and $A(t)$ from the introduction of §2 and observe that

$$
\begin{aligned}
J_1(u) &= \int_0^1 \{P(t) - A(t)\}dt \\
&\geq \begin{cases} \inf\limits_{t\in(0,1)} P(t) - A(t) & \text{if } \inf\limits_{t\in(0,1)} P(t) - A(t) < 0, \\ 0 & \text{if } \inf\limits_{t\in(0,1)} P(t) - A(t) \geq 0. \end{cases}
\end{aligned}
\tag{2.13}
$$

Therefore it follows from the minimality of J_1 at u that

$$P(t) - A(t) = \inf_{t\in(0,1)} P(t) - A(t),$$

i.e. almost every level set of u is a solution to Problem P1.

Conversely, if $v \in BV(\Omega)$ is any nonnegative function and if $P(t)$ and $A(t)$ are the perimeter and volume of the level sets $\Omega_t = \{x \in \Omega \,|\, v(x) > t\}$, then an inspection of (2.13) reveals that $J_1(v) \geq J_1(\chi_D)$, where D is a solution to Problem P1. This proves the second part of the theorem. ∎

Remark. The following nonuniqueness result for Problem P1 was kindly pointed out to me by S.Luckhaus. Let Ω consist of a large ball in $I\!R^2$ with a long "nose" of width 2 and with circular tip of radius 1, i.e.

$$\Omega := B_R(0) \cup \{(x, y) \in I\!R^2 \,|\, |y| < 1,\, 0 \leq x \leq 2R\} \cup B_1((2R, 0)).$$

Then for any $a \in [0, R,]$ the following set D_a is a solution of Problem P1:

$$D_a := B_R(0) \cup \{(x,y) \in I\!R^2 \mid |y| < 1,\ 0 \le x \le R + a\} \cup B_1((R+a,0)).$$

Consequently a solution of Problem P2 can have the sets D_a as level sets. Therefore Problem P2 can also have more than one solution.

Now it is relatively easy to establish an existence result for Problem P1.

Theorem 7. *Let $\Omega \in I\!R^N$ be bounded with Lipschitz continuous boundary $\partial\Omega$. Then Problem P1 has a solution D.*

For the proof we show that Problem P2 has a solution and apply Theorem 6. The functional $J_1(v)$ is bounded below by $-A(\Omega)$, so that there is a minimizing sequence $\{v_n\}_{n \in I\!N}$ which is bounded in $BV(\Omega)$. But then, after passing to a subsequence, v_n converges strongly in $L^1(\Omega)$ to a limit v_∞ as $n \to \infty$. Since the BV-norm is lower semicontinuous in $L^1(\Omega)$ we may conclude that v_∞ is a solution to Problem P2. Finally one can apply Theorem 6 to see that Problem P1 has a solution. ∎

What can be said about the regularity of ∂D for a solution D of Problem P1? The part of ∂D which lies inside Ω is a surface of prescribed constant mean curvature $1/(N-1)$, and the part of ∂D which lies on $\partial\Omega$ is as smooth as $\partial\Omega$. Moreover, even for smooth $\partial\Omega$, ∂D is of class $C^{1,1}$, but in general not of class C^2. This follows from regularity results for $(N-1)$-dimensional obstacle problems, see e.g. [KS,Lu,GMT] and the references therein.

With this amount of knowledge in mind, one can sometimes construct solutions of Problem P1. If $\Omega \subset I\!R^2$, then there are three alternatives possible: a) $D = \emptyset$, b) $D = \Omega$ and c) D is nontrivial and $\partial D \cap \Omega$ consists of circular arcs of radius 1 which touch the boundary of Ω tangentially. Let us study these possibilities if $\Omega = (-a,a) \times (-a,a)$ is a square in $I\!R^2$.

a) If $D = \emptyset$ then $P(D) - A(D) = 0$.
b) If $D = \Omega$ then $P(D) - A(D) = 4a(2-a) =: \gamma_1(a)$. The function $\gamma_1(a)$ is zero at $a = 2$ and at $a = 0$.
c) If D has circular arcs in the corner of the square (and if $a > 1$), then $P(D) - A(D) = \pi - 4(a-1)^2 =: \gamma_2(a)$. The function $\gamma_2(a)$ is zero at $a = 1 + \sqrt{\pi}/2 \approx 1.89$.

So for $0 < a < 1+\sqrt{\pi}/2 \approx 1.89$ Problem P1 has $D = \emptyset$ as a solution. For $0 < a < \sqrt{\pi} \approx 1.77$ this was also predicted by Theorem 3. If $a = 1+\sqrt{\pi}/2$, then Problem P1 has two solutions, $D = \emptyset$ and the one that consists of the square with its corners rounded off. For $a > 1 + \sqrt{\pi}/2$ we have to observe that $\gamma_2(a) - \gamma_1(a) < 0$ so that alternative c) is always preferred over alternative b). Therefore the square provides an example where the solution of Problem P1 (and of Problem P2) is nontrivial in the sense that $\partial D \cap \Omega \neq \emptyset$.

In a similar fashion one can show the following result. It is based on the observation that for the cost functional $P(D) - A(D)$ it pays off to make the corners round.

Theorem 8. *Let $\Omega \subset I\!R^2$ be bounded and convex and let $D_1 := \bigcup\{B_1(x_0) \,|\, B_1(x_0) \subset \Omega\}$ be the subset of Ω which is generated by "sweeping" Ω with balls of radius 1 from the inside. Then $D = D_1$ or $D = \emptyset$ is a solution of Problem P1.*

The proof can first be done for convex polygons, then for arbitrary convex domains, because the solutions of Problem P2 and hence of Problem P1 are monotone with respect to Ω. I leave the details to the reader. ∎

In a final remark I want to point out that Problem P2 is the limit case of another class of elliptic problems, namely so-called dead core problems.

$$\Delta_p w = H(w) \qquad \text{in } \Omega, \tag{6.3}$$

$$w = 1 \qquad \text{on } \partial\Omega, \tag{6.4}$$

where H is the Heaviside function, i.e. $H(t) = 1$ for $t > 0$,and $H(t) = 0$ for $t \leq 0$. For $p = 2$, these problems are well understood, see e.g. [BSS,C]. If Ω is large enough, the solution will have a nontrivial dead core $D = \{x \in \Omega \,|\, w(x) = 0\}$. For convex Ω the dead core is convex [K3]. If we set $v = 1 - w$, then for $1 < p < \infty$ the solution v is characterized by

$$J_p(v) \leq J_p(\tilde{v}) = \int_\Omega \{\frac{1}{p}|\nabla \tilde{v}|^p - \tilde{v}\}dx$$

for any nonnegative $\tilde{v} \in W_0^{1,p}(\Omega)$ with $||\tilde{v}||_{L^\infty(\Omega)} \leq 1$. Notice that Problem P2 is the limiting case of this class of variational problems. Furthermore, the effect that $D \neq \emptyset$ only if Ω is large can be expected from the corresponding result for $p = 2$.

Bibliography

[A] Aronsson,G., Construction of singular solutions to the p-harmonic equation and its limit equation for p=∞. manuscripta math. **56** (1986) p.135-158

[AT] Aronsson,G., G.Talenti, Estimating the integral of a function in terms of a distribution function of its gradient. Bolletino U. M. I. (5) **18**-B (1981) p.885-894.

[B] Bandle,C. *Isoperimetric inequalities and applications.* Pitman, (1980) Boston

[Ba] Bandle,C., On isoperimetric gradient bounds for Poisson problems and problems of torsional creep. J. Appl. Math. Phys. (ZAMP) **30** (1979) p.713-715

[BSS] Bandle,C., R.P.Sperb and I.Stakgold, Diffusion and reaction with monotone kinetics. Nonlinear Analysis **8** (1984) p.321-333

[C] Caffarelli,L.A., Compactness methods in free boundary problems. Comm. Partial Differ. Equations **5** (1980) p.427-448

[D] De Giorgi,E., Sulla proprieta isoperimetrica dell'ipersfera. Atti Accad. Lincei Mem. Cl. Fis. Mat. Natur. Sez Ia, (8) **5** (1958) p.33-44

[Gi] Giarusso,E., Su una classe di equazioni non lineari ellitiche. Ric. Mat. **31** (1982) p.245-257

[G] Giusti,E., *Minimal surfaces and functions of bounded variation.* Birkhäuser Verlag (1984) Basel

[GMT] Gonzales,E., U.Massari, I.Tamanini, On the regularity of boundaries of sets minimizing perimeter with a volume constraint. Indiana Un. Math. J. **32** (1983) p.25-37

[GKR] Goodman,J., R. Kohn, L.Reyna, Numerical study of a relaxed variational problem from optimal design. Computer Methods in Appl. Math. Engin. **57** (1986) p.107-127

[K1] Kawohl, B., *Rearrangements and convexity of level sets in PDE.* Springer Lecture Notes in Math. **1150** (1985) Heidelberg

[K2] Kawohl,B., Comparison results between linear and quasilinear elliptic boundary value problems. J. Math. Anal. Appl. **142** (1989) 162-169

[K3] Kawohl,B., When are solutions to nonlinear elliptic boundary value problems convex? Comm. Partial Differ. Equations **10** (1985) p.1213-1225

[K4] Kawohl,B., On a family of torsional creep problems. J. reine angew. Math. **410** (1990) p.1-22

[KSW] Kawohl,B., J.Stara, G.Wittum, Analysis and numerical studies of a shape design problem. Arch. Ration. Mech. Anal., to appear

[Ke] Keller,J., Plate failure under pressure. SIAM Review **22** (1980) p.227-228

[KS] Kinderlehrer,D.,G.Stampacchia, *An introduction to variational inequalities and their applications.* Academic Press, (1980) London

[L] Langenbach,A. Verallgemeinerte und exakte Lösungen des Problems der elastisch-plastischen Torsion von Stäben. Math. Nachr. **28** (1964) p.219-234

[Li] Lindquist,P. Stability for the solutions of $\mathrm{div}(|\nabla u|^{p-2}\nabla u) = f$ with varying p. J. Math. Anal. Appl. **127** (1987) p.93-102

[Lu] Luckhaus,S., Partial boundary regularity for the parametric capillarity problem in arbitrary space dimension. Habilitationsschrift, (1987) Heidelberg

[M1] Modica,L. The gradient theory of phase transitions and the minimal interface criterion. Arch. Ration. Mech. Anal. **98** (1987) p.123-142

[M2] Modica,L. Gradient theory of phase transitions with boundary contact energy. Preprint 776, Bonn 1986.

[PP] Payne,L.E., G.A.Philippin, Some applications of the maximum principle in the problem of torsional creep. SIAM J. Appl. Math. **33** (1977) p.446-455

[P] Philippin,G.A. A mimumum principle for the problem of torsional creep. J. Math. Anal. Appl. **68** (1979) p.526-535

[S] Sakaguchi,S. Concavity properties of solutions to some degenerate quasilinear elliptic Dirichlet problems. Ann. Sc. Norm. Sup. Pisa Ser. IV **14** (1987) p.404-421

[St] Steiner,J. Sur le maximum et le minimum de figures dans le plan, sur le sphere et dans espace en general I,II. J. reine angew. Math. **24** (1842) p.93-152, 189-250

[Ta] Talenti,G. Some estimates of solutions to Monge-Ampere equations in dimension two. Ann. Scuola Norm. Sup. Pisa Cl. Sci. IV **8** (1981) p.183-230

[Ti] Ting,T.W., Elastic plastic torsion of simply connected cylindrical bars. Indiana Univ. Math. J. **20** (1971) p.1047-1076

[T] Tolksdorf,P., Regularity for a more general class of quasilinear elliptic equations. J. Differ. Equations **51** (1984) p.126-150

Bernhard Kawohl
SFB 123, Universität Heidelberg
Im Ncucnheimer Feld 294
D 6900 Heidelberg, Germany

J P RAYMOND[1]

Nonconvex problems in calculus of variations

The study of non-convex variational problems has been motivated by Optimal Control problems [24, 25, 7,8] and more recently by problems related to phase transformations in solids and in fluids [14, 15, 16, 4, 5] (see also [9, 17] and their references for phase transitions in Elastic Crystals and [26] for propagation of phase boundaries).

Let us consider the following abstract minimization problem:

(P) $$\text{Inf } \{ J(u), u \in U \},$$

where J is a functional from U into $\mathbf{R} \cup \{+\infty\}$ and U is a separated topological space.

To prove the existence of solutions for (P), the so-called "Direct Method" consists in showing that J is bounded from below on U, J is coercive and lower semi-continuous on U. In Calculus of Variations, we often have:

(1) $$J(u) = \int_\Omega f(x,u(x),Du(x))\, dx, \quad u \in u_0+W_0^{1,q}(\Omega;\mathbf{R}^m),$$

$1 \le q < +\infty$, $u_0 \in W^{1,q}(\Omega;\mathbf{R}^m)$, Ω is a bounded open subset of $\mathbf{R}^n$ and U takes values in $\mathbf{R}^m$. In this case, U is the space $u_0+W_0^{1,q}(\Omega;\mathbf{R}^m)$ endowed with its weak topology.

Suppose that the function f in (1) satisfies some regularity assumptions and some growth conditions, it is well known that J is weakly lower semicontinuous (shortly w. l. s. c.) on $W^{1,q}$ if and only if f(x,u,.) is quasiconvex for all $x \in \Omega$ and all $u \in \mathbf{R}^m$ (cf [10, chap. 4]). In the case when m = 1 or n = 1, the condition "f(x,u,.) is quasiconvex" becomes "f(x,u,.) is convex".

<u>RELAXED PROBLEM</u>. In the problems related to phase deformations and in some optimal control problems, the functional J in problem (P) is bounded from below and coercive on the space U, but not l. s. c. on U. In this case, the minimizing sequences admit cluster points which are solutions of a new variational problem called "relaxed problem associated with (P)". Let us give a classical example. We set: $U = W_0^{1,4}(0,1)$, $J(u) = \int_0^1 \{(u'(x)^2 - 1)^2 + u(x)^2\}\, dx$.

We define the minimizing sequence $(u_n)_{n\in N^*}$ in the following way:

$$u_n \in W_0^{1,\infty}(0,1),\ u_n'(x) \in \{-1, 1\} \text{ a. e. on } [0,1],$$

$$u_n((2i-1)/(2n)) = 1/(2n),\ u_n(i/n) = 0 \text{ if } i \in \{1, \dots, n\}.$$

---

[1]Université Paul Sabatier, Laboratoire d'Analyse Numérique 31062 Toulouse Cedex - France -

We can easily see that the sequence $(u_n)_n$ converges weakly to the zero function which is the unique solution of the relaxed problem defined by:

(PR) $$\text{Inf } \{\bar{J}(u) \mid u \in W_0^{1,4}(0,1)\},$$

$$\bar{J}(u) = \int_0^1 \{ [(u'^2 - 1)_+]^2 + u^2 \} \, dx,$$

where $(v)_+ = \text{Max}(0, v)$.

There are different ways to define a relaxed problem, for this, we refer to [11, 13]. Here, we define the relaxed problem associated with (P) by:

(PR) $$\text{Inf } \{\bar{J}(u), u \in U\},$$

where $\bar{J}$ is the lower semicontinuous regularization of J [13]. The interest of the relaxed problem is contained in the following therorem.

Theorem 1 . If J is coercive and bounded from below on U, then:

(i) the relaxed problem (PR) admits solutions,

(ii) $\text{Inf } \{J(u) \mid u \in U\} = \text{Inf } \{\bar{J}(u) \mid u \in U\}$,

(iii) if (P) admits solution, every solution of (P) is a solution of (PR),

(iv) u is a solution of (P) if and only if u is a solution of (PR) and $J(u) = \bar{J}(u)$.

The solutions of (P) are called "exact solutions of (P)" as opposed to the solutions of (PR) which are often called "generalized solutions of (P)".

Most existence results of solutions for problems of Calculus of Variations with lack of lower semicontinuity (i. e. when J is bounded from below and coercive but not l. s. c.) are obtained either by constructive methods (often using the Lyapounov convexity theorem) or in consequence of analytic conditions on the integrands of the considered problems. Here, we shall not speak about the first method which was initially developed by NEUSTADT, OLECH and CESARI [24, 25, 7] (See also [6, 30] for recent developments). The second method, first developed by I. EKELAND [12], admit in dimension 1 (n = 1) an interesting geometrical interpretation. We shall describe it as "the geometrical method".

THE GEOMETRICAL METHOD FOR SCALAR PROBLEMS IN DIMENSION 1.

Now, we set:

$$J(u) = \int_a^b f(x,u(x),u'(x)) \, dx,$$

$$U = \{u \in W^{1,q}(a,b) \mid u(a) = \alpha, u(b) = \beta \}.$$

The relaxed problem associated with (P) is defined with:

$$\bar{J}(u) = \int_a^b f^{**}(x,u(x),u'(x)) \, dx,$$

where $f^{**}(x,u,.)$ is the biconjugate function of $f(x,u,.)$, (this result is in [13]). The condition (iv) in the theorem 1 is now:

(2) u is a solution of (P) if and only if u is a solution of (PR) and $f(x,u(x),u'(x)) = f^{**}(x,u(x),u'(x))$ a. e. on [a, b].

Let us suppose that f^{**} is locally Lipschitz. With classical assumptions on f (see [27]) we can avoid Lavrentiev phenomena ([3, 8]) and we can prove that, for every solution u of (PR), there exists an absolute continuous function p such that:

(3)
$$u'(x) \in \partial_p H(x,u(x),p(x)), \text{ a. e. on } [a,b],$$

$$p'(x) \in f^{**}_u(x,u(x),u'(x)), \text{ a. e. on } [a,b],$$

where

$$H(x,u,p) = \text{Sup}\ \{\mathcal{H}(x,u,p,v) \mid v \in \mathbf{R}\},$$

$$\mathcal{H}(x,u,p,v) = pv - f(x,u,v).$$

(f^{**}_u denotes the partial generalized gradient of f^{**} with respect to u). The lack of convexity of $f(x,u,.)$ is related to the occurence of multisingularities for $\mathcal{H}(x,u,p,.)$. More precisely, if $f(x,u,.)$ is not convex at $\bar{v}$, then the function $\mathcal{H}(x,u,p,.)$, with $p = f^{**}_v(x,u,\bar{v})$, reaches its maximum in at least two critical points (see fig. 1 below). In this case, for fixed $\bar{v}$, "$p = f^{**}_v(x,u,\bar{v})$" is a local equation of a submanifold in the phase space. So we set:

$$O = \{\ (x,u,v) \in [a,b]x\mathbf{R}^2 \mid f(x,u,v) \neq f^{**}(x,u,v)\}$$

and

$$\Sigma = \{(x,u,p) \in [a,b]x\mathbf{R}^2 \mid \exists\ v \in \mathbf{R} \text{ such that } (x,u,v) \in O \text{ and } p = f^{**}_v(x,u,v)\}.$$

For every solution u of (PR), we consider the trajectory $(x,u(x),p(x))_{x\in[a,b]}$ where p is defined by (3) and C_u is the set of contact between this trajectory and Σ :

$$C_u = \{x \in [a,b] \mid (x,u(x),p(x)) \in \Sigma\ \}.$$

We can easily see that condition (2) is equivalent to:

$$\text{Meas } C_u = 0.$$

In other words, studying existence of solutions for (P) is the same as studying contacts between the trajectories of (PR) and the submanifold Σ. If the conditions of tangent contact between Σ and the trajectories of (PR) can never be fulfilled, then every solution of (PR) is a solution of (P). This gives the following condition for the existence of solutions (see [27, prop 5.4]).

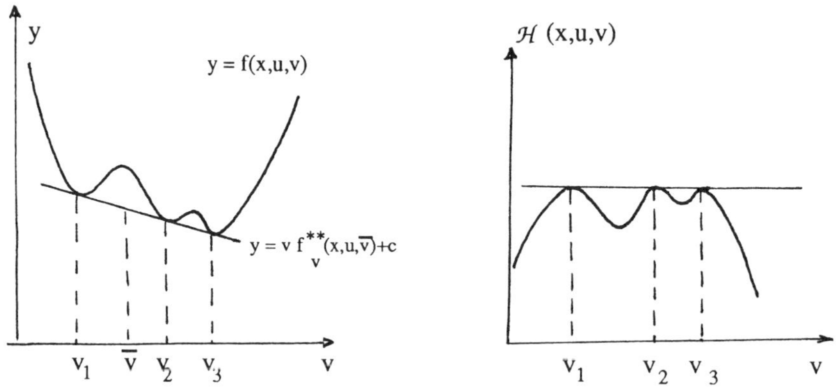

Figure 1

Theorem 2. If f^{**} is of class C^2 on O and if, for every $(x,u,v) \in O$, we have:

$$(4) \qquad f^{**}_u(x,u,v) - f^{**}_{vx}(x,u,v) - f^{**}_{uv}(x,u,v)\, v \neq 0,$$

then (P) admits solutions.

(In [1], AUBERT and TAHRAOUI studied the case when $f(x,u,v) = g(x,v)+h(x,u)$. See also [20], for new results of this kind).

This geometrical approach can also be used to obtain necessary and sufficient conditions of existence for scalar or vector problems (see [12, 18, 27, 28]).

PROBLEMS IN HIGHER DIMENSION. Now we consider the problem defined by:

$$J(u) = \int_\Omega \{g(Du(x)) + h(x,u(x))\}\, dx,$$

$u \in U = u_0 + W_0^{1,q}(\Omega;\mathbf{R}^m)$, $u_0 \in W^{1,q}(\Omega;\mathbf{R}^m)$, U is endowed with its weak topology.

The function g is not necessarily convex nor quasiconvex. We suppose that J is coercive and bounded from below on U. Here, the condition (4) is given by:

$$(5) \qquad h_u(x,u) \neq 0, \text{ for every } x \in \Omega \text{ and every } u \in \mathbf{R}^m.$$

Does the condition (5) ensure the existence of solutions for the problem (P) ? This question remains open in the general case. An answer has been given only in the case when g^{**} satisfies the following affinity condition:

(6) the set $O = \{v \in \mathbf{R}^{nm} \mid g(v) \neq g^{**}(v)\}$ is an at most countable union of sets $(O_i)_{i \in I}$ and g^{**} is affine on each set O_i.

Remark. This condition is always satisfies if $m = n = 1$.

We make the following assumptions on g and h.

(7) $\qquad g \in C^2(\mathbf{R}^{nm})$ and for every $v \in \mathbf{R}^{nm}$, we have:

$$c_1 (|v|^q - 1) \leq g(v) \leq c_2 (|v|^q + 1),$$

$$| g_{vv}(v) | \leq c_3 (|v|^{q-2} + 1), \quad q \geq 2,$$

where c_1, c_2, c_3 are positive constants and $g_{vv}(v)$ denotes the hessian matrix of g at v.

(8) $h \in C^2(\Omega x \mathbf{R}^m)$ and for every $x \in \Omega$ and every $u \in \mathbf{R}^m$,

$$a_1(x) - b_1|u|^q \leq h(x,u) \leq a_2(x) + b_2|u|^q,$$

$$| h_u(x,u) | \leq b_3 |u|^r + a_3(x),$$

$$| h_{ux}(x,u) | \leq b_4 |u|^{r/q'} + a_4(x),$$

$$| h_{uu}(x,u) | \leq b_5 |u|^{r(q-2)/q} + a_5(x),$$

with $b_1 < c(q,\Omega,m)^{-q} c_1$, $c(q,\Omega,m)$ being the constant in Poincaré inequality in $W_0^{1,q}(\Omega;\mathbf{R}^m)$, $a_1 \in L^1(\Omega)$, $a_2 \in L^1(\Omega)$, $r = n\,q / (n-q)$ if $q < n$ and r is any number greater than 1, if $q \geq n$, $a_3 \in L^1(\Omega)$, $a_4 \in L^{q'}(\Omega)$, $a_5 \in L^{q/(q-2)}(\Omega)$, $q' = q/(q-1)$.

Theorem 3. If assumptions (5), (6), (7), (8) are satisfied, then (P) admits solutions.

A similar theorem has been given in the scalar case (m = 1) and under more restrictive assumptions in [29, 21]. Here we claim that this result is also true in the vector case (m > 1). To prove this result, we consider the convexified problem:

$(\bar{P})$ $$\text{Inf} \{ \bar{J}(u) \mid u \in U\},$$

$$\bar{J}(u) = \int_\Omega \{g^{**}(Du(x)) + h(x,u(x))\}\, dx.$$

Remark. If m = 1, $\bar{J}$ is the lower semicontinuous regularization of J on U, but this result is not still true if m > 1.

In fact, we prove that condition (5) is sufficient so that every solution of $(\bar{P})$ be a solution of (P). For this, we need a regularity result on adjoint solutions of $(\bar{P})$.

Definition. The adjoint solution associated with a solution u of $(\bar{P})$ is the mapping p from Ω into $\mathbf{R}^{mn}$ defined by:

$$p(x) = g_v^{**}(Du(x)).$$

($g_v^{**}(v)$ denotes the gradient of g^{**} at v).

Remark: Since u is a solution of $(\bar{P})$, one has:

$$\operatorname{div} p(x) = h_u(x,u(x)).$$

So, a priori the function p belongs to the space :

$$\{p \in L^{q'}(\Omega;\mathbf{R}^{mn}) \mid \operatorname{div} p \in L^{q'}(\Omega;\mathbf{R}^m)\}.$$

But we can prove the following regularity result.

Theorem 4. If g and h satisfy (7) and (8), then the adjoint solutions of $(\bar{P})$ belong to $W^{1,nq'/(n-2+q')}_{loc}(\Omega;\mathbf{R}^{mn})$ and the solutions of $(\bar{P})$ belong to $W^{1,nq/(n-2)}$ if $n > 2$ and to $W^{1,r}_{loc}$ for every $r < +\infty$ if $n = 2$.

Similar results have been obtained for other problems in [31]. Let us give some ideas about the proof.

1st step. We first prove that g^{**} satisfies the following ellipticity inequality:

$$(9) \qquad (g_v^{**}(v_1) - g_v^{**}(v_2))(v_1 - v_2) \geq$$

$$\geq c\,(|g_v^{**}(v_1)| + |g_v^{**}(v_2)| + 2)^{(2-q)/(q-1)}(g_v^{**}(v_1) - g_v^{**}(v_2))^2,$$

for every $v_1 \in \mathbf{R}^{nm}$ and every $v_2 \in \mathbf{R}^{nm}$, c is a positive constant.

2nd step. With the Nirenberg's difference quotient method and with inequality (9), we prove that the adjoint solutions of $(\bar{P})$ belong to $W^{1,nq'/(n-1+q')}_{loc}(\Omega;\mathbf{R}^{nm})$.

With the Sobolev's imbedding theorem, we get $p \in L^{nq'/(n-q')}_{loc}$ and by a duality argument $u \in W^{1,nq/(n-q')}_{loc}$.

3rd step. Starting from $p \in L^{nq'/(n-q')}_{loc}$ and applying the step 2, we can improve the regularity result on p.

We prove the theorem by repeating this process and by passing through the limit.

Sketch of the proof of theorem 3.

Let u be a solution of $(\bar{P})$ and $A_i = \{x \in \Omega \mid Du(x) \in O_i\}$ (O_i is a set on which g^{**} is affine, see (6)). If x belongs to A_i, then $p(x) = c_i \in \mathbf{R}^{nm}$. With the regularity result of theorem 4, we can assert:

$$h_u(x,u(x)) = \operatorname{div} p(x) = 0 \quad \text{a.e. on } A_i.$$

(In fact, we only need $p \in W^{1,1}_{loc}$). If Meas $A_i \neq 0$, we get a contradiction with condition (5).

So, Meas $A_i = 0$ for all $i \in I$, and we have proved that $J(u) = \bar{J}(u)$. Thus u is a solution of (P).

Comments. Other existence or inexistence results are given in [2, 4, 19, 20, 21, 22, 23] and this suggest us two open questions.

(i) If condition (6) is not satisfied, can we give sufficient conditions on h in order that (P) admit solutions ?

(ii) In [19], P. MARCELLINI gave a problem (P) for which the condition (6) is not satisfied, $h \equiv 0$, u_0 belongs to $C^1(\bar{\Omega})$ and $g(Du_0(x)) \neq g^{**}(Du_0(x))$ for every $x \in \bar{\Omega}$. He proved that, in this case, (P) does not admit any solution (see also [4] for other similar results).

If (6) is not satisfied and if $h \equiv 0$, can we give necessary and sufficient conditions (or at least sufficient conditions) on the boundary data (u_0 and $\partial\Omega$) in order that (P) admit a solution ?
In particular, does the conditions:

$$u_0 \in C^1(\bar{\Omega}), \ \ g(Du_0(x)) = g^{**}(Du_0(x)) \ \text{ for all } x \in \bar{\Omega},$$

ensure (in the scalar case) the existence of solutions for (P) ?

References

[1] AUBERT, G., TAHRAOUI, R., Théorèmes d'existence pour des problèmes du Calcul des Variations. J. Diff. Eq., t. 33, n° 1, 1979, p. 1-15.

[2] AUBERT, G., TAHRAOUI, R., Théorèmes d'existence en optimisation non convexe. Applicable Analysis, 18, 1984, p. 75-100.

[3] BALL, J.M., MIZEL, V.J., One dimensional variational problems whose minimizers do not satisfy the Euler-Lagrange equations. Arch. Rat. Mech. An., Vol. 90, 1985, p. 325-388.

[4] BAUMAN, P., PHILLIPS, D., A Nonconvex Variational Problem Related to Change of Phase. Appl. Math. Optim., 21, 1990, p. 113-138.

[5] CARR J., GURTIN, M.E., SLEMROD, M., Structured phase transitions on a finite interval. Arch. Rat. Mech. An. 86, 1984, p. 317-351.

[6] CELLINA, A., COLOMBO, G., On a Classical Problem of the Calculus of Variations without Convexity Assumptions. An. I.H.P., An. non lin., Vol. 7, n° 2, 1990, p. 97-107.

[7] CESARI, L., An Existence Theorem without Convexity Conditions. SIAM J. Control 12, 1974, p. 319-331.

[8] CESARI, L., Optimization, Theory and Applications. Springer, 1983.

[9] CHIPOT, M., KINDERLEHRER, D., Equilibrium Configurations of Crystals. Arch. Rat. Mech. An. 103, 1988, p. 237-277.

[10] DACOROGNA, B., Direct Methods in the Calculus of Variations. Springer, 1989.

[11] EKELAND, I., Sur le Contrôle Optimal de Systèmes gouvernés par des Equations Elliptiques. J. Funct. An., Vol. 9, 1972, p. 1-62.

[12] EKELAND, I., Discontinuités de Champs Hamiltoniens et Existence de Solutions Optimales en Calcul des Variations. Publications Math. de l'I.H.E.S., Vol. 47, 1977, p. 5-32.

[13] EKELAND, I., TEMAM, R., Analyse Convexe et problèmes variationnels. Dunod, Paris, 1974.

[14] ERICKSEN, J.L., Equilibrium of bars. J. of Elasticity, Vol. 5, 1975, p. 191-201.

[15] FOSDICK, R.L., MAC SITHIGH, G., Helical Shear of an Elastic, Circular Tube with a Non-convex Stored Energy. Arch. Rat. Mech. An., Vol. 84, 1983, p. 31-53.

[16] JAMES, R.D., Co-existent Phases in the On-Dimensional Static Theory of Elastic Bars. Arch. Rat. Mech. An., Vol. 72, 1979, p. 99-140.

[17] JAMES, R.D., KINDERLEHRER, D., Theory of Diffusionless Phase Transitions in "PDEs and Continuum Models of Phase Transitions", M. RASCLE, D. SERRE, M. SLEMROD Eds., 1988, L. N. in Physics 344, Springer.

[18] KLOK, F., On non existence of minimizing solutions of variational problems. J. Math. Anal. Applic., Vol. 114, 1986, p. 129-155.

[19] MARCELLINI, P., A relation between existence of minima for non convex integrals and uniqueness for non strictly convex integrals of the Calculus of Variations, Springer, Lect. Not. Math., Vol. 979, 1983, p. 216-232.

[20] MARCELLINI, P., Non convex Integrals of the Calculus of Variations. To appear. Springer, Lect. Not. Math.

[21] MASCOLO, E., Some remarks on Non-Convex Problems, in Material Instabilities in Continuum Mechanics, J.M. BALL Ed. Clarendon Press, Oxford, 1988, p. 269-286.

[22] MASCOLO, E., SCHIANCHI, R., Existence theorems for non convex problems. J. Math. Pures Appl., Vol. 62, 1983, p. 349-359.

[23] MASCOLO, E., SCHIANCHI, R., Existence theorems in the Calculus of Variations. J. Diff. Eq., Vol. 67, 1987, p. 185-198.

[24] NEUSTADT, L.W., The existence of Optimal Control in the Absence of Convexity Conditions. J. Math. An. Applic., 1963, p. 110-117.

[25] OLECH, C., Integrals of set-valued functions and Optimal control problems. Colloque sur la théorie du Contrôle Optimal, Vander Louvain, Belgique, 1970, p. 109-125.

[26] PEGO, R.L., Phase transitions in one dimensional non linear viscoelasticity: admissibility and stability. Arch. Rat. Mech. An. 97, 1987, p. 353-394.

[27] RAYMOND, J.P., Conditions nécessaires et suffisantes d'existence de solutions en Calcul des Variations. Annales de l'I.H.P., An. non lin., Vol. 4, 1987, p. 169-202.

[28] RAYMOND, J.P., Champs Hamiltoniens, relaxation et existence de Solutions en Calcul des Variations. J. Diff. Eq. Vol. 70, 1987, p. 226-274.

[29] RAYMOND, J.P., Théorème d'Existence pour des Problèmes Variationnels non Convexes. Proc. Royal Soc. Ed., 107 A, 1987, p. 43-64.

[30] RAYMOND, J.P., Existence Theorems in Optimal Control Problems without Convexity Assumptions. To appear in J.O.T.A..

[31] RAYMOND, J.P., Théorème de régularité locale pour des systèmes elliptiques dégénérés et des problèmes non différentiables. Ann. Fac. Sc. Toulouse, Vol. 9, 1988, p. 381-412.

HYPERBOLIC PROBLEMS - EXACT CONTROLABILITY

F CONRAD

Stabilization of Euler–Bernoulli beams with nonlinear dissipation on the boundary

Université de Nancy II and U.R.A. CNRS 750,
B.P. 239, 54506 VANDOEUVRE-lès-NANCY, France.

INTRODUCTION.

Research in control and stabilization of distributed parameter systems has been considerably stimulated in recent years by an increasing number of practical problems concerning large space structures, flexible robots ... Of particular interest is the design and analysis of feedback controls which stabilize, or which increase the degree of stability, of a system.

Due to the complexity of real world flexible systems, research in stabilization has largely focused on single components of flexible structures such as strings, membranes, beams, plates. Even for such simple models, it is of interest to consider nonlinear feedback, since, from a practical point of view, control acts usually in a nonlinear way, by means of boosters, nonlinear friction... We refer for instance to [WA-CHE] for a justification of nonlinear energy dissipation conditions in modern material science and technology.

It is the purpose of this lecture to present a study on stabilization of a nonhomogeneous Euler-Bernoulli beam, which is controlled at one end by a shear force or a bending moment (inputs), defined as nonlinear functions of the transversal and angular velocities at the same end (observations).

Since this workshop is primarily concerned with exact controllability, it may be useful to recall that there is a close connection between (boundary) stabilizability and exact (boundary) controllability, at least for linear problems. On the one hand, an argument due to RUSSEL [RU1], [RU2] shows that for a linear time reversible system, existence of boundary controls, which uniformly stabilize the system, implies exact boundary controllability, after a large enough delay. On the other hand, following LIONS [LI], one can always associate with an exactly controllable system a regulator, i. e. a problem of (linear quadratic) optimal control with an infinite horizon, which can usually be realized as a linear feedback in the following way: one obtains for the regulator the corresponding optimality conditions; this system of equations can be uncoupled; this leads to feedback formulas giving exponential decay in some appropriate norm.

Therefore, roughly speaking, one can say that exact controllability is (more or less !) equivalent to uniform exponential stabilization for linear reversible systems.

Exact controllability is an attractive property, but, from a practical point of view, stabilization is the required property, at least for two simple reasons:

- even if there exists a control which drives the system to rest in finite time, generally, when perturbation occurs on the data, the system no longer goes to zero, even in infinite time.
- even if exact controllability implies the existence of a stabilizing feedback by the general method indicated above, the realization of the feedback is usually not completely explicit, or simple.

In this lecture, we follow a direct method for obtaining stabilization. It consists in considering a priori (suitably !) given feedbacks, which dissipate "enough" energy. This technique has been used very exhaustively. First we present a short review of some typical results on nonlinear boundary stabilization. Then we study the asymptotic stability of a model of nonhomogeneous Euler-Bernoulli beam. Finally, we give energy estimates for this model in the case of sublinear feedback.

2. NONLINEAR BOUNDARY STABILIZATION OF SOME ELASTIC SYSTEMS.

Our presentation is far from being complete. We just mention typical papers which have some connection with our problem which is studied in Sections 3 and 4.

2.1. One-dimensional wave equation.

In [WA-CHE] the following model for the motion of a cable or a string, is considered:

$$\begin{cases} y_{tt} - y_{xx} = 0 \text{ in } (0, 1) \\ y(1, t) = 0 \\ y_x(0, t) \in f(y_t(0, t)) \end{cases}$$

where $f(0) = [-\alpha, +\alpha]$, and f is a monotone regular function elsewhere.

Using the method of characteristics and nonlinear semi-group theory, the authors study the asymptotic behaviour of the solutions of an equivalent system where the unknowns are $\frac{1}{2}(y_t - y_x)$ and $\frac{1}{2}(y_t + y_x)$. In fact, the ω-limit set of the dynamical system is explicitely characterized. A solution may either enter the ω-limit set in finite time, or circle around it indefinitely, without enterring it. Asymptotic rates of decay to the ω-limit set are given when $f \pm \alpha$ is linear, sublinear, or superlinear. When $\alpha = 0$, the ω-limit set reduces to zero.

2.2 Wave equation in arbitrary dimension ($N = 1, 2, 3$).

In [LA1], [LA2] the following extension of the above model is considered:

$$\begin{cases} y_{tt} - \Delta y = 0 \text{ in } \Omega \\ y = 0 \text{ on } \Gamma_0 \\ \dfrac{\partial y}{\partial n} + \gamma(y_t) \ni 0 \text{ on } \Gamma_1 \end{cases}$$

where n is the unit normal pointing outwards, γ is a monotone graph such that $0 \in \gamma(0)$, and $\partial\Omega = \Gamma_0 \cup \Gamma_1$.

If $\gamma(0) = \{0\}$, it is proved that the system is strongly asymptotically stable in the topology of $H^1(\Omega) \times L^2(\Omega)$, provided $(x - x_0).\, n \le 0$ on Γ_0, with some growth condition on γ if $N > 1$. In the case $\gamma(0) \neq \{0\}$, provided moreover $(x - x_0).\, n \ge 0$ on Γ_1, the energy of all limit solutions is bounded uniformly w.r.t. initial conditions, by C_Ω diam($\gamma(0)$), where C_Ω depends only on the geometry of Ω. This gives an estimate of the ω-limit set.

Remark. The case where the Neumann condition $\frac{\partial y}{\partial n} + \gamma(y_t) \ni 0$ on Γ_1 is replaced by a Dirichlet condition: $y + \gamma(F y_t) \ni 0$ on Γ_1, with $F = \frac{\partial A^{-1}}{\partial n}$, and $A y = \Delta y$ on $H_0^1(\Omega) \cap H^2(\Omega)$, is also considered in [LA2]. If $\gamma(0) = \{0\}$, then strong stability in $L^2(\Omega) \times H^{-1}(\Omega)$ is proved, with the same assumptions as in the case of Neumann condition.

Estimates of the decay have been obtained in a similar framework. For instance in [ZU], the following system is considered:

$$\begin{cases} y_{tt} - \Delta y = 0 \text{ in } \Omega \\ y = 0 \text{ on } \Gamma_0 = \{x \in \partial\Omega : (x - x_0).\ n \le 0\} \\ \dfrac{\partial y}{\partial n} = - (x - x_0).n\ g(y_t) \text{ on } \Gamma_1 = \{x \in \partial\Omega : (x - x_0).n > 0\} \end{cases}$$

where int $(\Gamma_0) \neq \emptyset$, and g is a nondecreasing continuous function such that $g(0) = 0$.

Assuming $C_1 \inf\{|s|, |s|^p\} \le |g(s)| \le C_2 |s|$, with $p \ge 1$, the following estimates for the energy $E(t) = \frac{1}{2} \int_\Omega (|\nabla y|^2 + |y_t|^2)\, dx$, are obtained:

(i) if $p = 1$, $E(t) \le M\, E(0)\, e^{-\delta t}$, where M and δ are positive constants

(ii) if $p > 1$, $E(t) \le M\, E(0) \left(\frac{1}{1+\lambda t}\right)^{\frac{2}{p-1}}$, where $M > 0$ is constant, and $\lambda > 0$ depends on E(0).

Remark 1. If int $(\Gamma_0) = \emptyset$ (i. e. Ω is star-shaped w.r.t. x_0), similar decay estimates have been obtained for the modified energy (in order to retain a norm):

$E_\alpha(t) = E(t) + \frac{\alpha}{2} \int_{\Gamma_1} (x-x_0).n\ |y|^2\, d\sigma$, with the modified feedback law:

$\frac{\partial y}{\partial n} = - \alpha\, (x - x_0).n\ y - (x - x_0).n\ g(y_t)$ on Γ_1, provided $\alpha > 0$ is not too large.

Remark 2. The preceding results are valid for a semi-linear wave equation: $y_{tt} - \Delta y + f(y) = 0$ in Ω, with all other conditions kept unchanged, provided $f(s).s \ge 0$, and f satisfies adequate growth conditions [KO-ZU].

2.3. Beam equation.

In [SLE] the following model of Euler-Bernoulli beam controlled by a truncated feedback is considered:

$$\begin{cases} u_{tt} + u_{xxxx} = 0 \text{ in } (0, L) \\ u(0, t) = u_x(0, t) = 0 \\ u_{xx}(L, t) = - u_{xtt}(L, t) + f(t) \\ u_{xxx}(L, t) = u_{tt}(L, t) \end{cases}$$

where $f(t) = \begin{cases} - u_{xt}(L, t) \text{ if } |u_{xt}(L, t)| < r \\ - r \operatorname{sign}(u_{xt}(L, t)) \text{ if } |u_{xt}(L, t)| \geq r \end{cases}$

The above problem is well-posed for $(u, u_t, u_t(L, t), u_{xt}(L, t))$ in $H^2(0,L) \times L^2(0,L) \times \mathbb{R}^2$, and asymptotic stability is proved, at least if $L = 1$.

An interesting feature of the model is that the feedback law is dynamic, thus leading to a hybrid system. Hybrid systems become very popular, in particular on account of the development of flexible robotics.

2.4. Plates with nonlinear dissipation on the boundary.

In [LA3] the following Kirchhoff plate model, with velocity feedback in the shear force and bending moment on the boundary, is considered:

$$\begin{cases} u_{tt} + \Delta^2 u = 0 \text{ in } \Omega \\ u = \dfrac{\partial u}{\partial n} = 0 \quad \text{on } \Gamma_0 \\ \Delta u + (1-\mu)B_1 u + K_1 \gamma_1 \left(\dfrac{\partial u_t}{\partial n}\right) \ni 0 \text{ on } \Gamma_1 \\ \dfrac{\partial}{\partial n}\Delta u + (1-\mu)\dfrac{\partial}{\partial \tau}B_2 u - K_2 \gamma_2(u_t) \ni 0 \text{ on } \Gamma_1 \end{cases}$$

where $K_1 \geq 0$, $K_2 \geq 0$, $n = (n_1, n_2)$ is the normal vector, $\tau = (-n_2, n_1)$ is the tangent vector, $B_1 = 2 n_1 . n_2 \dfrac{\partial^2}{\partial x \partial y} - n_1^2 \dfrac{\partial^2}{\partial y^2} - n_2^2 \dfrac{\partial^2}{\partial x^2}$, $B_2 = (n_1^2 - n_2^2) \dfrac{\partial^2}{\partial x \partial y} +$

$n_1 . n_2 \left(\dfrac{\partial^2}{\partial y^2} - \dfrac{\partial^2}{\partial x^2}\right)$, γ_i are maximal monotone graphs such that $\gamma_i(0) = \{0\}$, and $0 < \mu < 1$ (in the physical situation, $\mu \in (0, 1/2)$ is the Poisson's ratio).

When $K_1 = 0$, $K_2 > 0$ (dissipation occurs only through shear forces), strong asymptotic stability in $H^2(\Omega) \times L^2(\Omega)$ is obtained with the "usual" geometric conditions $(x - x_0). n \leq 0$ on Γ_0, $(x - x_0). n \geq 0$ on Γ_1. The case $K_2 > 0$, $K_1 > 0$ works also provided $(x - x_0). n > 0$ on Γ_1 and γ_1 has at most polynomial growth. If $\gamma_i(0) \neq \{0\}$, an estimate of the ω-limit set is given.

The more difficult case of control by bending moment only ($K_1 > 0, K_2 = 0$), has also been considered [LA3], [LA4]. Roughly speaking, the preceding result is true, provided dissipation acts on the whole boundary (the condition $\frac{\partial u}{\partial n} = 0$ on Γ_0 is removed), and the curvature of the boundary of Ω is small.

Remark. Other dissipative nonlinear boundary conditions of lower order, involving more complicated boundary operators, as in the case of the wave equation (Section 2.2), have been considered in [LA2].

Concerning nonlinear boundary stabilization of plates, and especially estimates for the decay of the elastic energy, the monograph by LAGNESE [LA] contains a large number of results. In the particular case of the preceding Kirchhoff model, with control by force only, one typical result is as follows. Let E(t) be the elastic energy of the system:

$$E(t) = \frac{1}{2} \int_{\Omega} [u_t^2 + u_{xx}^2 + u_{yy}^2 + 2\mu\, u_{xx}u_{yy} + 2(1-\mu)u_{xy}^2]\, dx\, dy \quad (\mu \in (0, 1/2)).$$

Assume the geometric conditions: $(x - x_0).\, n \leq 0$ on Γ_0, $(x - x_0).\, n \geq 0$ on Γ_1, with $\Gamma_0 \neq \emptyset$ (or, if $\Gamma_0 = \emptyset$, suitable modification of the feedback law), and $cl(\Gamma_0) \cap cl(\Gamma_1) = \emptyset$. Assume $K_2\, \gamma_2(\xi)$ is replaced by $(x - x_0).\, n\, g(\xi)$, where g is a nondecreasing regular function such that, roughly speaking, $g(\xi)$ behaves like $|\xi|^{p-1}\xi$ near 0, with $p \geq 1$ (and g satisfies some appropriate growth conditions). Then:

(i) if $p = 1$, $E(t) \leq M\, E(0)\, e^{-\delta t}$

(ii)if $p > 1$, $E(t) \leq M\, E(0) \left(\frac{1}{1+\lambda t}\right)^{\frac{2}{p-1}}$.

This result is the one obtained for the wave equation [ZU], and in fact the technique of obtention is an adaptation of that case, too.

From the standpoint of asymptotic stability and decay estimates, in a nonlinear framework, many other models of beams or plates, for instance thermoelastic plates, the von Karman system... have been studied by LAGNESE and others, see the references in [LA].

All the models presented up to now involved equations with constant coefficients. There are comparatively much less results for nonhomogeneous wave, or beam, or plate, equations, at least for the stabilization problem. The case of serially connected Euler-

Bernoulli beams has been considered in [CHE-DE]. This system is clamped at the left end. The right end is controlled by a force and a moment which are linear dissipative functions of the transversal and angular velocities. Uniform exponential decay is proved when the force is active, provided the linear mass densities are nondecreasing and the flexural rigidities are nonincreasing. It is not clear whether these assumptions of monotonicity are really necessary, or not.

It is our purpose in the next two sections to show that for a similar model, with nonlinear boundary feedback, no such assumption is necessary to derive strong asymptotic stability. Likewise, for the obtention of estimates of the decay, the above-mentioned monotonicity assumptions just mentioned can be relaxed.

3. NONHOMOGENEOUS EULER-BERNOULLI BEAMS: STABILIZATION BY NONLINEAR BOUNDARY FEEDBACK.

This section reflects a joint work with M. PIERRE [CO-PI].

3.1. Statement of the problem.

We consider an Euler-Bernoulli beam with variable mass density and flexural rigidity. The beam is clamped at one end. The other end is controlled by a point force and a point bending moment which are assumed to be nonlinear functions of the observation. By observation we mean velocity and angular velocity of the transversal deflection at the end.

Let u(x, t) denote the transversal deflection at point x and time t. For a beam of unit length, with flexural rigidity a(x) and linear mass density b(x), we get the following equations:

Open-loop system

$$\begin{cases} b\, u_{tt} + (a\, u_{xx})_{xx} = 0 \text{ in } (0, 1) & \\ u(0, t) = u_x(0, t) = 0 & \text{(the beam is clamped at } x = 0) \\ -(a\, u_{xx})(1, t) = f_1(t) & \text{(moment)} \\ (a\, u_{xx})_x(1, t) = f_0(t) & \text{(force)} \\ u(x, 0) = u_0(x),\ u_t(x, 0) = u_1(x) & \end{cases} \qquad (1)$$

with $a \in L^\infty(0, 1)$; $b \in L^\infty(0, 1)$; $0 < m \leq a, b$. We refer to [RU3] where many mathematical models for the elastic beam are presented.

Closed-loop system

Extending the case of linear feedback [CHE-DE], we assume the controls f_0 and f_1 to be nonlinear nondecreasing functions of the observed velocities $u_t(1, t)$ and $u_{xt}(1, t)$, vanishing at zero. This implies dissipation of the elastic energy of the system. Even from a practical point of view, it may be of interest to consider nonlinear feedback involving some "abrupt" behaviour for small values of the argument, or some "saturating effect" for large values of the argument.

Thus, we consider in this papcr general feedback laws of the form:

$$f_0(t) \in \alpha(u_t(1, t))$$
$$f_1(t) \in \beta(u_{xt}(1, t))$$

where α and β are maximal monotone graphs in $\mathbb{R}^2$ such that $0 \in \alpha(0)$, $0 \in \beta(0)$. Dissipativity is induced by the fact that $\xi\, \alpha(\xi)$ and $\xi\, \beta(\xi) \subset \mathbb{R}_+$.

Therefore the system with boundary feedback control we study is the following:

$$\begin{cases} b\, u_{tt} + (a\, u_{xx})_{xx} = 0 \quad \text{in } (0, 1) \\ u(0, t) = u_x(0, t) = 0 \\ -(a\, u_{xx})(1, t) \in \beta\, (u_{xt}(1, t)) \\ (a\, u_{xx})_x(1, t) \in \alpha\, (u_t(1, t)) \end{cases} \tag{2}$$

Our formulation includes the case of serially connected beams, with regular (variable) mass density and flexural rigidity for each beam. The beams are strongly connected, with no force or bending moment applied at the internal nodes.The system is clamped at the left end, and control is exerted at the right end.

3.2. Wellposedness of Problem (2).

We show that Problem (2) defines a nonlinear semi-group of contractions generated by a maximal monotone operator , on the natural space of energy.

We set $H = L^2(0, 1)$, endowed with the norm $|v|_H = (\int_0^1 b\, v^2\, dx)^{1/2}$,

$V = \{u \in H^2(0, 1)\ ;\ u(0) = u_x(0) = 0\}$, endowed with the norm $|u|_V = (\int_0^1 a\, u_{xx}^2)^{1/2}$, and V x H, endowed with the product norm, is the "energy" space.

Dividing the equation in (2) by b and setting $v = u_t$, we write system (2) in the form:

$$\frac{d}{dt}\binom{u}{v} + A(u, v) \ni 0, \quad (u, v) \in V \times H \tag{3}$$

where the nonlinear operator A on V x H is defined by:

$\mathrm{Dom}(A) = \{ (u, v) \in V \times H;\ a\, u_{xx} \in H^2(0, 1);\ v \in V;\ v(1) \in \mathrm{Dom}(\alpha);$
$v_x(1) \in \mathrm{Dom}(\beta);\ -(a\, u_{xx})(1) \in \beta(v_x(1));\ (au_{xx})_x(1) \in \alpha(v(1)) \}$

and, if $(u, v) \in \mathrm{Dom}(A)$: $A(u, v) = \begin{pmatrix} -v \\ \frac{1}{b}(a\, u_{xx})_{xx} \end{pmatrix}$.

Theorem 1. A is a maximal monotone operator with dense domain. Moreover, $\forall (f,g) \in V \times H$, $(u, v) = (I + A)^{-1}(f, g)$ satisfies: $|u|_V^2 + |v|_V^2 + |a\, u_{xx}|_{H^2}^2 \leq C\, [\, |f|_V^2 + |g|_H^2\,]$.

Proof. The monotonicity is easy to set up, as well as the density property. To prove that $(u, v) + A(u, v) = (f, g) \in V \times H$, admits a (unique) solution $(u, v) \in V \times H$, we first establish the existence of a solution and the estimate, on a Yosida-regularization of the equation, then pass to the limit, in a way very similar to [BRE2]. We refer to [CO-PI] for all the details. Note that the estimate in Theorem 1 implies that the resolvent map is compact on V x H.

Now the evolution problem (2) makes sense in a standard way.

Theorem 2. (i) For any initial data $(u_0, u_1) \in V \times H$, Problem (3) admits a unique "mild solution". In fact, (3) defines a nonlinear semi-group of contractions S(t) on V x H.
(ii) If $(u_0, u_1) \in \mathrm{Dom}(A)$, then u is a strong solution of (2) or (3), i.e.: $(u, u_t) \in \mathrm{Dom}(A)$ a.e.(t); $t \in (0, +\infty) \to (u, u_t) \in V \times H$ is (a.e.) differentiable w.r.t. t, and (2) or (3) are satisfied a.e. in t.

Proof. For Problem (3), (i) and (ii) are direct consequences of Theorem 1 by standard results [BRE1, PA]. We just have to note that for (ii), $u_{tt} \in H$, $\frac{1}{b}(a\, u_{xx})_{xx} \in H$, so that we can multiply the equation of (3) by b to get (2).

3.3. Asymptotic behaviour.

Applying the invariance principle of LASALLE [DA-SLE], combined with elementary nonharmonic Fourier analysis in order to derive a uniqueness theorem, we obtain the strong asymptotic stability under "minimal" assumptions on the graphs α and β. In particular, stability is achieved with control, either by a shear force only, or by a bending moment only.

Let $(u, u_t) = S(t)(u_0, u_1)$ be the solution of (3) for initial data $(u_0, u_1) \in V \times H$.

Theorem 3. (active dissipation through shear force). Assume
(i) $\alpha(0) = \{0\}$, and $\forall\, \xi \neq 0, 0 \notin \alpha(\xi)$
(ii) $\beta(0) = \{0\}$, and $\forall\, \xi, 0 \in \beta(\xi) \Rightarrow \beta(\xi) = \{0\}$
Then, $\forall\ (u_0, u_1) \in V \times H$, $S(t)(u_0, u_1) \to 0$ in $V \times H$ as $t \nearrow \infty$.

Theorem 4. (active dissipation through bending moment). Assume
(i) $\beta(0) = \{0\}$, and $\forall\, \xi \neq 0, 0 \notin \beta(\xi)$
(ii) $\alpha(0) = \{0\}$, and $\forall\, \xi, 0 \in \alpha(\xi) \Rightarrow \alpha(\xi) = \{0\}$
Then, $\forall\ (u_0, u_1) \in V \times H$, $S(t)(u_0, u_1) \to 0$ in $V \times H$ as $t \nearrow \infty$.

Sketch of proofs. Again we refer to [CO-PI] for the details.

Step1. For any $(u_0, u_1) \in V \times H$, the elastic energy of a solution $(u, u_t) = S(t)(u_0, u_1)$ of (3) is given by:

$$E(t) = \frac{1}{2} \|(u(t), u_t(t))\|^2_{V\times H} = \frac{1}{2} \int_0^1 [b\, u_t^2(x, t) + a\, u_{xx}^2(x, t)]\, dx.$$

Then, if $(u_0, u_1) \in \mathrm{Dom}\,(A)$, we have the energy formula

$$E(t) - E(0) \in - \int_0^t [\ \alpha(u_t(1, s))u_t(1, s) + \beta(u_{xt}(1, s))u_{xt}(1, s)]\, ds \tag{4}$$

This formula is obtained by differentiation of E(t), and integration by parts, taking System (2) into account. The procedure is valid by the regularity of the data and Theorem 2.
In particular, E(t) is nonincreasing w.r.t. t. By the density of Dom (A) in $V \times H$, and the contraction property of the semi-group, the monotonicity holds also if $(u_0, u_1) \in V \times H$.
Thus $\frac{1}{2} \|(u, v)\|^2_{V\times H}$ is a Lyapunov function on $V \times H$.

Step 2. By density of Dom (A) in V x H and the contraction property of the semi-group, it is enough to prove Theorems 3 and 4 for $(u_0, u_1) \in$ Dom (A). Let $(u_0, u_1) \in$ Dom (A), and $\omega = \omega(u_0, u_1)$ be the ω-limit set of (u, u_t). This ω-limit set exists by compacity of the resolvent map (Theorem 1) and the fact that 0 is in the range of A [DA-SLE, PA]. Also $\omega \subset$ Dom (A) [BRE1]. In order to show that $(u, u_t) \to 0$, it is enough to prove that $\omega = \{0\}$.

Step 3. Let $(w_0, w_1) \in \omega \subset$ Dom (A), and $(w(t), w_t(t)) = S(t)(w_0, w_1) \in \omega$. It is enough to prove that $(w(t), w_t(t)) = 0, \forall\, t > 0$! Apply the invariance principle of LASALLE:

$$E(t) = \frac{1}{2}\|(w(t), w_t(t))\|^2_{VxH} = \frac{1}{2}\int_0^1 [b\, w_t^2(x, t) + a\, w_{xx}^2(x, t)]\, dx \quad \text{is constant w.r.t. t.}$$

By the energy formula (4) applied to (w_0, w_1), we deduce that

$$0 \in \int_0^t [\alpha(w_t(1, s))w_t(1, s) + \beta(w_{xt}(1, s))w_{xt}(1, s)]\, ds$$

which, since $\xi\, \alpha(\xi)$ and $\xi\, \beta(\xi) \subset R_+$, implies that

$0 \in \alpha(w_t(1, t))w_t(1, t)$, and $0 \in \beta(w_{xt}(1, t))w_{xt}(1, t)$, a.e. in t.

In other words, in order to get the strong asymptotic stability for (3), it is sufficient to establish the following "uniqueness" result:

Let w be a solution of the "overdetermined" system (5), (6):

$$\begin{cases} b\, w_{tt} + (a\, w_{xx})_{xx} = 0 \quad \text{in } (0, 1) \\ w(0, t) = w_x(0, t) = 0 \\ -(a\, w_{xx})(1, t) \in \beta\, (w_{xt}(1, t)) \\ (a\, w_{xx})_x(1, t) \in \alpha\, (w_t(1, t)) \\ (w_0, w_1) \in \omega(u_0, u_1) \subset \text{Dom (A)} \end{cases} \tag{5}$$

$$\begin{cases} 0 \in \alpha(w_t(1, t))w_t(1, t) \\ 0 \in \beta(w_{xt}(1, t))w_{xt}(1, t) \end{cases} \tag{6}$$

then necessarily $(w, w_t) \equiv 0$.

Step 4. Now we consider the specific case of Theorem 3.

From (6) and assumption (i) in Theorem 3, $w_t(1, t) = 0$, thus $\alpha(w_t(1, t)) = \{0\}$, a.e. t.

From (6) and assumption (ii) in Theorem 3 we get $\beta(w_{xt}(1, t)) = \{0\}$, a.e. t.

Thus w satisfies the following system:

$$\begin{cases} b\, w_{tt} + (a\, w_{xx})_{xx} = 0 \text{ in } (0, 1) \\ w(0, t) = w_x(0, t) = 0 \\ -(a\, w_{xx})(1, t) = 0 \\ (a\, w_{xx})_x(1, t) = 0 \\ (w_0, w_1) \in \text{Dom } (A) \end{cases} \tag{7}$$

with the subsidiary equation induced by (6) and assumption (i) in Theorem 3

$$w_t(1, t) = 0 \tag{8}$$

Step 5. Let $\lambda_n > 0$ (n = 1, 2,...) be the sequence of eigenvalues of the following problem:

$$\begin{cases} (a\, \phi_{xx})_{xx} = \lambda\, b\, \phi \text{ in } (0, 1) \\ \phi(0) = \phi_x(0) = 0 \\ (a\, \phi_{xx})(1) = (a\, \phi_{xx})_x(1) = 0 \end{cases} \tag{9}$$

We assume that the associated eigenvectors ϕ_n are normalized by $\|\phi_n\|_H^2 = \int_0^1 b\, \phi_n^2\, dx = 1.$

The solution of (7) is given by the following expansion:

$$w(t) = \sum_{n=1}^{n=\infty} [a_n \cos(\sqrt{\lambda_n}\, t) + b_n \sin(\sqrt{\lambda_n}\, t)]\, \frac{\phi_n}{\sqrt{\lambda_n}} \tag{10}$$

$$w_t(t) = \sum_{n=1}^{n=\infty} [b_n \cos(\sqrt{\lambda_n}\, t) - a_n \sin(\sqrt{\lambda_n}\, t)]\, \phi_n \tag{11}$$

where

$$a_n == \frac{1}{\sqrt{\lambda_n}} \int_0^1 a(x)\, w_{0xx}(x)\, \phi_{nxx}(x)\, dx; \quad b_n = \int_0^1 b(x)\, w_1(x)\, \phi_n(x)\, dx.$$

The series given by (10) (resp. (11)) converges in V (resp. H). Since (w_0, w_1) is in Dom (A), and in fact in Dom(A_0), these convergences can be improved.

Specificaly, $w_1 = \sum_{n=1}^{n=\infty} b_n \phi_n \in V$, thus $\sum_{n=1}^{n=\infty} \lambda_n b_n^2 < \infty$, since $|\phi_n|_V^2 = \lambda_n$ and

$$\sqrt{\lambda_n}\, b_n = \langle w_1, \frac{\phi_n}{\sqrt{\lambda_n}} \rangle_{V \times V'}$$

Similarly, $w_0 = \sum_{n=1}^{n=\infty} a_n \frac{\phi_n}{\sqrt{\lambda_n}} \in V$, thus $a_n \in l^2(\mathbb{N})$.

Moreover, since $(w_0, w_1) \in \mathrm{Dom}\,(A_0)$, $(aw_{0xx})_{xx} \in L^2(0, 1)$ and

$\sqrt{\lambda_n}\, a_n = \langle (aw_{0xx})_{xx} , \phi_n \rangle_{L^2xL^2}$.

Thus $\sum_{n=1}^{n=\infty} \lambda_n a_n^2 < \infty$.

As $\sqrt{\lambda_n}\, a_n$ and $\sqrt{\lambda_n}\, b_n \in l^2(\mathbb{N})$, the series (11) converges in V, uniformly w.r.t. t. By the continuity of the trace operator $v \to v(1)$ on V, the subsidiary equation (8), $w_t(1, t) = 0$ a.e. in t, rewrites as

$$\sum_{n=1}^{n=\infty} [b_n \phi_n(1) \cos (\sqrt{\lambda_n}\, t) - a_n \phi_n(1) \sin (\sqrt{\lambda_n}\, t)] = 0 \text{ a.e. } t > 0 \qquad (12)$$

where the above numerical series converges uniformly w.r.t. t.

Assume for a moment that the λ_n are all distinct. From the uniform convergence in (12), we deduce that the Fourier coefficients of the almost periodic function 0 must be precisely $b_n \phi_n (1)$ and $-a_n \phi_n (1)$. Thus $a_n \phi_n (1) = b_n \phi_n (1) = 0$.

Assume now that $\phi_n(1) \neq 0$, $\forall$ n (observability). Then $a_n = b_n = 0$ for any $n \geq 1$, which implies $(w,w_t) \equiv 0$.

The simplicity of the eigenvalues (λ_n all distinct) and the observability condition ($\phi_n (1) \neq 0$ for all n) are easy consequences of the following result [CO-PI]:

Lemma. Let $\phi \in H^2(0, 1)$, such that $a\, \phi_{xx} \in H^2(0, 1)\}$, be a solution of:

$$\begin{cases} (a\, \phi_{xx})_{xx} = \lambda\, b\, \phi \quad \text{in } (0, 1) \\ \phi(0) = \phi_x(0) = 0 \\ (a\, \phi_{xx})(1) = (a\, \phi_{xx})_x(1) = 0 \\ \phi(1) = 1 \end{cases}$$

Then $\phi \equiv 0$ on (0, 1).

Results on the simplicity of the eigenvalues for fourth order O. D. E. 's can be found for instance in [LE-NE, Section 4]. There, it is in fact asserted that simplicity is a quite general property of fourth order equations with regular coefficients, at least for equations in "variational" form. However, the precise statement we need here (nonregular coefficients and specific choice of boundary conditions) was not stated explicitely. Moreover, we need also the observability result. Therefore, we have adapted the method of proof of [LE-NE].

Then the proof of Theorem 3.3 is complete. Theorem 4 is established in a similar way (replace $\phi_n(1)$ by $\phi_{nx}(1)$ and $\phi(1)$ by $\phi_x(1)$).

Remark. A proof of Theorems 3 and 4 by means of standard multipliers leads to supplementary assumptions of monotonicity involving a and b.

4. ESTIMATES OF THE DECAY.

This section reflects a joint work with J. LEBLOND and J. P. MARMORAT. We consider Problem (2) with control by force only ($\beta = 0$), a less complicated case than the case of control by moment. We assume α is a monotone continuous sublinear function such that

$$\begin{cases} \alpha(\xi)\,\xi \geq a_1\,|\xi|^r + b_1\,|\xi|^2 & (H_1) \\ \alpha(\xi)\,\xi + c_1\,|\xi|^s \geq d_1\,|\alpha(\xi)|^2 & (H_2) \end{cases}$$

where $a_1, b_1, c_1, d_1 > 0$, and $0 < s \leq 2(r-1) \leq r \leq 2$. Typically, assumptions (H_1) and (H_2) are satisfied when $\alpha(\xi) = c\,|\xi|^{\varepsilon-1}\xi + d\,\xi \;\; \forall\, c, d > 0$, and $0 < \varepsilon \leq 1$.

Remark. The following fact has been proved in [CO-LE], in a somewhat abstract framework. Assume the feedback is (approximately) linear. Then uniform decay of the energy (that is, $E(t) \leq \varepsilon(t)\,E(0)$, with $\lim \varepsilon(t) = 0$ as $t \nearrow \infty$, uniformly w.r.t. $E(0)$) can be expected only if the following strong observability condition is satisfied: there exists $T > 0$ such that

$$|\phi_t(0)|_H^2 + |\phi(0)|_V^2 \leq M \int_0^T \phi_t^2(1, t)\,dt \tag{13}$$

where ϕ is the solution of the uncontrolled system

$$\begin{cases} b\,\phi_{tt} + (a\,\phi_{xx})_{xx} = 0 \text{ in } (0, 1) \\ \phi(0, t) = \phi_x(0, t) = 0 \\ (a\,\phi_{xx})(1, t) = (a\,\phi_{xx})_x(1, t) = 0 \\ \phi(0),\ \phi_t(0) \text{ are given} \end{cases}$$

Clearly, (13) is stronger than the uniqueness property needed in Section 3 in order to prove the asymptotic stability (Theorem 3). Therefore, it is expected that simplicity of the eigenvalues and the observability property established in Section 3 will not be sufficient here. In fact, some "gap" on the eigenvalues $\sqrt{\lambda_n}$ is necessary for (13) to hold. We do not know if such a property is always true for the general problems we study. We state only a result with monotonicity assumptions on a and b. However, these assumptions are weaker than those used in [CHE-DE] for stabilizing serially connected beams with linear feedback.

Recall that the elastic energy of a solution (u, u_t) of (2) is given by

$$E(t) = \frac{1}{2} \int_0^1 [b\, u_t^2(x, t) + a\, u_{xx}^2(x, t)]\, dx$$

<u>*Theorem 5*</u>. Let α be satisfy (H_1) and (H_2). Assume moreover that

(H_3) a and b are piecewise of class C^1 on (0, 1), xb and $\frac{x^3}{a}$ are nondecreasing on [0, 1]

(H_4) $\exists\ \varepsilon > 0$ such that $(x\,b)_x \geq \varepsilon\, b$, $(\frac{x^3}{a})_x \geq 2\varepsilon \frac{x^2}{a}$ a.e. on [0, 1]. Then

(i) if $r = s\ (= 2)$, $E(t) \leq M\, E(0)\, e^{-\delta t}$, where M and δ are positive constants.

(ii) if $r < s$, $E(t) \leq M\, E(0) \left(\frac{1}{1+\lambda\, t}\right)^{\frac{s}{r-s}}$, where $M > 0$ is constant and $\lambda > 0$ depends on E(0).

Remark 1. When $\alpha(\xi)$ behaves like $|\xi|^{\varepsilon-1}\xi$ for small ξ $(0 < \varepsilon \leq 1)$, the energy decays like $\left(\frac{1}{t}\right)^{\frac{2\varepsilon}{1-\varepsilon}}$. Control by linear feedback appears to be the best one ! Note also that by transforming ε into $\frac{1}{\varepsilon}$, our estimates are similar to those obtained in the superlinear case for waves or plates [LA], [ZU], in the case of constant coefficients.

Remark 2. Our result fits similar decay estimates obtained in a sublinear framework in [LA-LE] for the case of a beam equation which models the effect of stretching on bending. The coupled equations for longitudinal and transversal displacement are nonlinear, and general nonlinear feedbacks on velocity and rate of bending are applied at one end of the beam. This model is much more complicated as ours. However, the coefficients are assumed to be constant.

Sketch of proof.

We use multipliers. By density and the contraction property, it is enough to prove the result for regular initial data, provided the constants appearing in the estimates depend only on E(0). Let (u, u_t) be a solution of (2). In the sequel, the C_i denote various constants.

Step 1. We multiply bu_{tt} by xu_x, and integrate over (0, 1) x (0, T). We first integrate by parts w.r.t. t, then over each interval of regularity w.r.t. x. Using the monotonicity of xb we get

$$\int_0^T\int_0^1 bu_{tt}\, xu_x\, dxdt \geq \int_0^1 [xbu_xu_t]_0^T\, dx + \frac{1}{2}\int_0^T\int_0^1 (xb)_x\, u_t^2\, dxdt - \frac{1}{2}(xb)(1)\int_0^T u_t^2(1,t)\, dt \quad (14)$$

Step 2. Similarly, we multiply $(au_{xx})_{xx}$ by xu_x, and integrate by parts w.r.t. x over each interval of regularity. Using the monotonicity of $\frac{x^3}{a}$, we get

$$\int_0^T\int_0^1 (au_{xx})_{xx}\, xu_x\, dxdt \geq \int_0^T \alpha(u_t(1,t))u_x(1,t)dt + \int_0^T\int_0^1 au_{xx}^2 dx + \frac{1}{2}\int_0^T\int_0^1 (\frac{x}{a})_x\, (au_{xx})^2 dx \quad (15)$$

Step 3. We add (14) and (15) and take into account the equation of (2) to get

$$\frac{1}{2}\int_0^T\int_0^1 (xb)_x\, u_t^2\, dx\, dt + \int_0^T\int_0^1 (1 + \frac{1}{2}a\, (\frac{x}{a})_x)au_{xx}^2 dx \leq$$

$$- \int_0^1 [xbu_xu_t]_0^T dx - \int_0^T \alpha(u_t(1, t))\, u_x(1, t)dt + \frac{1}{2}(xb)(1)\int_0^T u_t^2(1, t)\, dt \quad (16)$$

Step 4. We recall the energy formula (4) $\int_0^t \alpha(u_s(1, s))u_s(1, s)\, ds = E(0) - E(t) \leq E(0).$

By this formula and (H_1), the last integral in (16) is bounded by $C_1E(0)$. By Young's inequality, combined with Poincaré's inequality for the term u_x, the absolute value of the first integral of the r.h.s. in (16) is bounded by C_2 [E(0) + E(T)] $\leq C_3$ E(0).

For the second integral of the r.h.s. in (16), we apply Young's inequality with suitable coefficients, and Poincaré's inequality, to get

$$\left| \int_0^T \alpha(u_t(1,t))\, u_x(1,t)\, dt \right| \le \frac{1}{2\mu} \int_0^T \alpha^2(u_t(1,t))\, dt + \frac{\mu}{2} C_4 \int_0^T \int_0^1 a u_{xx}^2\, dx \tag{17}$$

By (H_2) and Hölder's inequality, the first integral of the r.h.s. of (17) is bounded by $\frac{C_5}{2\mu}[E(0) - E(T)] + \frac{C_6}{2\mu} T^{\frac{r-s}{r}} \left(\int_0^T u_t^2(1,t) dt \right)^{\frac{s}{r}}$. Again by (H_1) this last integral can be bounded by $C_7\, [E(0) - E(T)]^{\frac{s}{r}}$. Injecting all these estimates into (16), we get

$$\frac{1}{2}\int_0^T \int_0^1 (xb)_x\, u_t^2\, dx\, dt + \int_0^T \int_0^1 \left(1 + \frac{1}{2} a \left(\frac{x}{a}\right)_x\right) a u_{xx}^2\, dx \le$$

$$\left(C_1 + C_3 + \frac{C_5}{2\mu}\right) E(0) + \frac{C_8}{2\mu} T^{\frac{r-s}{r}} [E(0)]^{\frac{s}{r}} + \frac{\mu}{2} C_4 \int_0^T \int_0^1 a u_{xx}^2 dx \tag{18}$$

Now, we choose $\mu\, C_4 \le \varepsilon$. Assumption (H_4) applied to (18) implies

$$\varepsilon\, T\, E(T) \le \varepsilon \int_0^T E(t)\, dt \le C_1^* E(0) + C_2^* T^{\frac{r-s}{r}} [E(0)]^{\frac{s}{r}} \tag{19}$$

Step 5. For $s = r = 2$, the exponential uniform decay follows immediately from (19) by taking T large enough. The case $s < r$ is a little more involved. We refer to [CO-LE] for the derivation of estimate (ii) in Theorem 5 from (19).

Finally, we would like to remark that though our monotonicity assumptions are weaker than those of [CHE-DE], they look still quite artificial and could certainly be relaxed.

REFERENCES

[[BRE1] H. BREZIS, Opérateurs Maximaux Monotones et Semi-groupes de Contraction dans les Espaces de Hilbert, North-Holland (1973).

[BRE2] H. BREZIS, Monotonicity Methods in Hilbert Spaces and Some Applications to Nonlinear Partial Differential Equations, in Contributions to Nonlinear Functional Analysis, ed. E. H. Zarantonello, Ac. Press (1971) pp. 101-157.

[CHE-DE] G. CHEN, M. DELFOUR, A. M. KRALL, and G. PAYRE, Modeling, stabilization and control of serially connected beams, SIAM J. Contr. Opt. 25, 3 (1987) pp. 526-546.

[CO-LE] F. CONRAD, J. LEBLOND, J. P. MARMORAT, Stabilization of second order evolution equations by unbounded nonlinear feedback, IFAC Symp. Contr. of Distributed Parameter Systems, Perpignan, june 1989 (Proc. to appear).

[CO-PI] F. CONRAD, M. PIERRE, Stabilization of Euler-Bernoulli beam by nonlinear boundary feedback, INRIA Report # 1235 (1990).

[DA-SLE] C. M. DAFERMOS, M. SLEMROD, Asymptotic behaviour of nonlinear contraction semi-groups, J. Funct. Anal. 13 (1973) pp. 97-106.

[KO-ZU] V. KOMORNIK, E. ZUAZUA, A direct method for the boundary stabilization of the wave equation, J. Math. Pures et Appl. 69, 1 (1990) pp. 33-54.

[LA] J. E. LAGNESE, Boundary Stabilization of Thin Plates (to appear).

[LA-LE] J. E. LAGNESE, G. LEUGERING, Uniform stabilization of a nonlinear beam by nonlinear boundary feedback (to appear).

[LA1] I. LASIECKA, Asymptotic behaviour of the solutions to wave equation with nonlinear damping on the boundary, Proc. Conf. Anal. Opt. Systèmes, Antibes (1988) pp. 472-483.

[LA2] I. LASIECKA, Stabilization of wave and plate-like equations with nonlinear dissipation on the boundary, J. Diff. Eqs. 79, 2 (1989) pp. 340-381.

[LA3] I. LASIECKA, Asymptotic behavior of solutions to plate equations with nonlinear dissipation occuring through shear forces and bending moments, Appl. Math. Opt. 21 (1990) pp. 167-189.

[LA4] I. LASIECKA, Asymptotic behaviour of the solutions of the Kirchhoff plate with nonlinear dissipation in the bending moment, Proc. IFIP Conf. Clermont-Ferrand (june 1988), in Control of Boundaries and Stabilization, J. Simon Ed., Lect Notes in Contr. and Inf. Sci. 125 (1989) pp. 168-176.

[LE-NE] W. LEIGHTON, Z. NEHARI, On the oscillation of solutions of self-adjoint linear differential equations of the fourth order, Trans. Amer. Math. Soc. 89 (1958) pp. 325-377.

[LI] J. L. LIONS, Exact controllability, stabilization, and perturbations for distributed systems, SIAM Review 30 (1988) pp. 1-68.

[PA] A. PAZY, Initial value problems for nonlinear differential equations in Banach spaces, Proc. Sem. Coll. de France 84, 5 (1982) pp. 154-172.

[RU1] D. L. RUSSEL, A unified boundary controllability theory for hyperbolic and parabolic partial differential equations, Studies in Appl. Math. 52 (1973) pp. 189-211.

[RU2] D. L. RUSSEL, Controllability and stabilizability theory for linear partial differential equations. Recent progress and open questions, SIAM Review 20 (1978) pp. 639-739.

[RU3] D. L. RUSSEL, Mathematical models for the elastic beams and their control theoretic implications, in Semi-groups, Theory and Applications, Eds. H. Brezis, M. G. Crandall, F. Kappel, Longman (1985) pp. 177-216.

[SLE] M. SLEMROD, Feedback stabilization of a linear control in Hilbert space with an a priori bounded control, Math. Control Signals Systems 2 (1988) pp. 265-285.

[WA-CHE] H. K. WANG, G. CHEN, Asymptotic behaviour of solutions of the one dimensional wave equation with a nonlinear boundary stabilizer, SIAM J. Contr. Opt. 27, 4 (1989) pp. 758-775.

[ZU] E. ZUAZUA, Uniform stabilization of the wave equation by nonlinear boundary feedback, SIAM J. Contr. Opt. 28, 2 (1990) pp. 466-477.

P GRISVARD

Stiff eigenvalue problems in solid mechanics

Summary : This work is about the behavior of the eigenmodes of an inhomogeneous body which has two components of different materials glued together. It is assumed that one of the materials is very light and "soft" as compared with the other. Mathematically this means that we look at what happens when the corresponding density and Lamé coefficient tend to zero. We reconstruct the physical situation with the help of asymptotic expansions.

Contents :

I. Junctions between components of same dimension

1. Introduction and Notation: The simplest case of a vibrating system having two components of different materials glued together is governed by a wave equation whith piecewise constant coefficients. Its eigenmodes are the eigenvalues of the variationnal problem which follows.

We consider an open bounded susbet Ω of $\mathbb{R}^n$ as in the following figure :

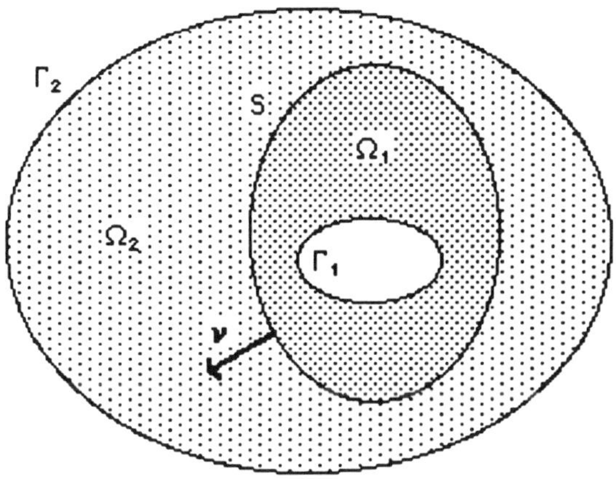

Consequently Ω is the union of two open sets Ω_1 (interior set: 1=i) and Ω_2 (exterior set : 2=e) and of their common boundary S. The interesting case is when S **is not empty.** The remaining part of the boundary of Ω_i is denoted Γ_i, i=1,2. One has $\Gamma = \partial\Omega = \Gamma_1 \cup \Gamma_2$. It will be convenient, but not necessary, to assume that Γ_1 and Γ_2 are not empty.

We introduce a real number $\varepsilon > 0$ and an integer m=1 or 2.

We are going to consider the variationnal problem defined by the following items :

The basic space is $H_\varepsilon = L^2(\Omega)$ equipped with the scalar product

$$b_\varepsilon(u,v) = \int_\Omega \rho_\varepsilon uv \, dx$$

$$\text{where } \rho_\varepsilon(x) = 1 \text{ for } x \in \Omega_1 \text{ and } \rho_\varepsilon(x) = \varepsilon \text{ for } x \in \Omega_2,$$

the energy-space V is $H^1{}_0(\Omega)$ and the energy-form is

$$a_\varepsilon(u,v) = \int_\Omega \sigma_\varepsilon \nabla u.\nabla v \, dx$$

$$\text{where } \sigma_\varepsilon(x) = 1 \text{ for } x \in \Omega_1 \text{ and } \sigma_\varepsilon(x) = \varepsilon^m \text{ for } x \in \Omega_2.$$

This form is continuous bilinear and coercive on $V \times V$.

The eigenvalues $\lambda \in \mathbb{R}$ and the corresponding eigenvectors $w \in V$, $w \neq 0$ under consideration are such that

$$(1) \qquad \forall v \in V: \quad a_\varepsilon(w,v) = \lambda b_\varepsilon(w,v).$$

2. Scope : We are going to investigate the dependance of these eigenelements in ε going to zero.

Since the injection of V into H_ε is compact and the form a_ε is symmetric and coercive, the eigenvectors define an orthonormal basis of H_ε for every $\varepsilon > 0$ and the corresponding eigenvalues are real and have finite multiplicities. We can number them in increasing order $\lambda_k(\varepsilon)$, $k=1,2,\ldots$ They depend continuously in $\varepsilon > 0$ (c.f. §3 below). Our scope is to understand their behavior as $\varepsilon \to 0$.

We shall denote by :

$$\sigma(-\Delta_D;\Omega_i) \text{ and } \sigma(-\Delta_N;\Omega_i)$$

respectively the spectra of the operator $-\Delta$ in the domain Ω_i under Dirichlet boundary condition on Γ_i but with

Dirichlet condition, **resp.** Neumann condition

on S (the subscript N actually refers to a mixed problem).

With the help of classical results on analytic perturbations of linear operators we are going to prove the following result where we take into account only simple eigenvalues :

Theorem 1 : We assume

$m=1$,

$\lambda_0 \in \sigma(-\Delta_D;\Omega_2) \cup \sigma(-\Delta_N;\Omega_1)$,

$\lambda_0 \notin \sigma(-\Delta_D;\Omega_2) \cap \sigma(-\Delta_N;\Omega_1)$.

We assume in addition that λ_0 is simple. Then problem (1) has an eigenvalue $\lambda(\varepsilon)$ and a corereesponding unitary eigenvector $w(\varepsilon)$ in H_1 which depend analytically on ε in a *neighborhood of* $\varepsilon=0$ *and such that*

$$\lambda(\varepsilon) \longrightarrow \lambda_0$$

when $\varepsilon \longrightarrow 0$.

It is therefore possible to compute $\lambda(\varepsilon)$ and $w(\varepsilon)$ with the help of the so-called "ansatz" method. This means looking for $\lambda(\varepsilon)$ and $w(\varepsilon)$ of the following form

$$\lambda(\varepsilon) = \lambda_0+\varepsilon\lambda_1+\varepsilon^2\lambda_2+\ldots$$

$$w(\varepsilon) = w_0 + \varepsilon w_1 + \varepsilon^2 w_2 +\ldots$$

as indicated in Lions (1988)

This is known and proved in Sanchez-Palencia (1980) and Sanchez-Hubert and Sanchez-Palencia (1989).

The second case where $m=2$ is not considered in loc.cit. (see however Chap.7 §6 in loc.sit. for a problem of a similar kind , meaning different powers in σ_ε and ρ_ε). There is only one family of eigenvalues "coming from" $\sigma(-\Delta_D;\Omega_2)$ and none "coming from" $\sigma(-\Delta_N;\Omega_1)$:

Theorem 2 : *We assume*

$m=2$,

$\lambda_0 \in \sigma(-\Delta_D;\Omega_2)$

We assume in addition that λ_0 *is simple*. *Then problem* (1) *has an eigenvalue* $\lambda(\varepsilon)$ *and a corresponding unitary eigenvector* $w(\varepsilon)$ *in* H_1 *which depends analytically on* ε *in a neighborhood of* $\varepsilon=0$ *and such that*

$$\lambda(\varepsilon)/\varepsilon \longrightarrow \lambda_0$$

when $\varepsilon \longrightarrow 0$.

Here it is possible to compute $\lambda(\varepsilon)$ assuming a priori the following form

$$\lambda(\varepsilon) = \varepsilon\{\lambda_0+\varepsilon\lambda_1+\varepsilon^2\lambda_2+\ldots\}$$

and $w(\varepsilon)$ as above.

We claim that there does not exist a second family of eigenvalues coming from $\sigma(-\Delta_N;\Omega_1)$. However we shall prove this claim only for a one dimensionnal problem similar to the above. In this problem we fix two positive real numbers a and b and we set $\Omega=]-a;b[$, $\Omega_1=]-a;0[$, $\Omega_2=]0;b[$ and $S=\{0\}$ with

$$b_\varepsilon(u,v) = \int_\Omega \rho_\varepsilon uv\,dx = \int_{-a}^0 uv\,dx + \varepsilon\int_0^b uv\,dx$$

and

$$a_\varepsilon(u,v) = \int_\Omega \sigma_\varepsilon u'v'\,dx = \int_{-a}^{0} u'v'\,dx + \varepsilon^m \int_0^b u'v'\,dx$$

Theorem 3 : We assume

$n=1$, $m=2$.

The only eigenvalues $\lambda(\varepsilon)$ of problem (1) which have a limit as $\varepsilon \to 0$ are those of Theorem 2.

It is reasonnable to conjecture that the result remains in the general case where $n \geq 1$, $m=2$.

3. Self-adjoint setting : It is a natural idea to consider the inverse of the unbounded operator A_ε defined by the variationnal problem (1). This leads to consider symmetric compact operators A_ε^{-1} in the space H_ε which unfortunately depends on ε. This does not allow one to apply the usual techniques of perturbations of eigenvalues.

This is why we shall define an operator similar to A_ε^{-1} in a fixed space.

Let us go back to (1). Written as partial differential equations with boundary conditions, (1) means that $w \in H^1{}_0(\Omega)$ and

$$-\mathrm{div}(\sigma_\varepsilon \nabla w) = \lambda \rho_\varepsilon w.$$

Define $\varphi = w\sqrt{\rho_\varepsilon}$. It is equivalent to claim that $\varphi/\sqrt{\rho_\varepsilon} \in H^1{}_0(\Omega)$ and that

$$-\mathrm{div}(\sigma_\varepsilon \nabla[\varphi/\sqrt{\rho_\varepsilon}])/\sqrt{\rho_\varepsilon} = \lambda\varphi.$$

We are thus led to consider the operator

$$L_\varepsilon : \varphi \mapsto -\mathrm{div}(\sigma_\varepsilon \nabla[\varphi/\sqrt{\rho_\varepsilon}])/\sqrt{\rho_\varepsilon}$$

in the space $H_1 = L^2(\Omega)$ and with domain

$$D(L_\varepsilon) = \{\varphi \in L^2(\Omega)\,;\, \varphi/\sqrt{\rho_\varepsilon} \in H^1{}_0(\Omega), \mathrm{div}(\sigma_\varepsilon \nabla[\varphi/\sqrt{\rho_\varepsilon}]) \in L^2(\Omega)\}.$$

This is the self-adjoint operator with compact inverse associated to the bilinear form

$$\varphi,\psi \mapsto l_\varepsilon(\varphi,\psi) = \int_\Omega \sigma_\varepsilon \nabla[\varphi/\sqrt{\rho_\varepsilon}]\nabla[\psi/\sqrt{\rho_\varepsilon}]\,dx$$

on the space $V_\varepsilon = \{\varphi \in L^2(\Omega)\,;\, \varphi/\sqrt{\rho_\varepsilon} \in H^1{}_0(\Omega)\}$.

We have now to study the behavior of the eigenvalues and the eigenvectors of L_ε in dependance of ε.

Let us observe that in this setting everything depends on $\sqrt{\varepsilon}$ instead of ε.

For every $f \in L^2(\Omega)$ there exists a unique $u(\varepsilon) \in V_\varepsilon$ solution of

$$\forall v \in V_\varepsilon : \quad l_\varepsilon(u(\varepsilon), v) = (f, v).$$

The inverse of L_ε is

$$S_\varepsilon : f \mapsto u(\varepsilon),$$

where $u(\varepsilon)$ is solution of

$$u(\varepsilon)/\sqrt{\rho_\varepsilon} \in H^1{}_0(\Omega), \ -\mathrm{div}(\sigma_\varepsilon \nabla[u(\varepsilon)/\sqrt{\rho_\varepsilon}]) = f\sqrt{\rho_\varepsilon}.$$

Denote by $u_i(\varepsilon)$ the restriction of $u(\varepsilon)$ to Ω_i. The above problem is equivalent to the transmission problem (the conditions on S express the continuity of $u(\varepsilon)/\sqrt{\rho_\varepsilon}$ and of $\sigma_\varepsilon \partial[u(\varepsilon)/\sqrt{\rho_\varepsilon}]/\partial\nu$ through S where ν is the unitary outward normal on Ω_1):

$$u_i(\varepsilon) \in H^1(\Omega_i), \ i=1,2,$$

$$u_i(\varepsilon) = 0 \text{ on } \Gamma_i, \ i=1,2,$$

$$-\Delta u_1(\varepsilon) = f_1 \text{ on } \Omega_1,$$

$$-\varepsilon^{m-1} \Delta u_2(\varepsilon) = f_2 \text{ on } \Omega_2,$$

$$\varepsilon^{1/2} u_1(\varepsilon) = u_2(\varepsilon) \text{ on } S,$$

$$\partial u_1(\varepsilon)/\partial\nu = \varepsilon^{m-1/2}(\partial u_2(\varepsilon)/\partial\nu) \text{ on } S.$$

This will be reduced to a pseudo-differential problem on S. For this purpose we introduce G_i and P_i the Green and Poisson operators of the Dirichlet problem on Ω_i. In other words

$$u = G_i f + P_i \varphi$$

is the solution of the problem

$$u \in H^1(\Omega_i),$$

$$-\Delta u = f \text{ in } \Omega_i,$$

$$u = 0 \text{ on } \Gamma_i,$$

$$u = \varphi \text{ on } S.$$

If one considers

$$\varphi(\varepsilon) = u_1(\varepsilon)|_S = \varepsilon^{-1/2} u_2(\varepsilon)|_S$$

as **auxiliary unknown** one gets

$$(2) \qquad u_1(\varepsilon) = G_1 f_1 + P_1 \varphi(\varepsilon), \ u_2(\varepsilon) = \varepsilon^{1-m} G_2 f_2 + \varepsilon^{1/2} P_2 \varphi(\varepsilon).$$

and

$$(3) \qquad \varepsilon^{m-1/2}(\partial/\partial\nu)\{\varepsilon^{1-m}G_2f_2 + \varepsilon^{1/2}P_2\varphi(\varepsilon)\} = (\partial/\partial\nu)\{G_1f_1 + P_1\varphi(\varepsilon)\} \text{ on } S.$$

In order to simplify we introduce the following notation

$$T_i\varphi = (\partial/\partial\nu)P_i\varphi|_S.$$

This defines a linear continuous operator T_i from $H^{1/2}(S)$ into $H^{-1/2}(S)$ and

$$g_i = (\partial/\partial\nu)G_if_i|_S \in H^{-1/2}(S).$$

The equation (3) may be rewritten as

$$(4) \quad (T_1-\varepsilon^mT_2)\varphi(\varepsilon) = -g_1+\varepsilon^{1/2}g_2.$$

We can now prove the

Proposition 4 : The function $\varepsilon \mapsto S_\varepsilon$ is analytic in $\varepsilon>0$ with values in $\mathcal{L}(H_1)$.

Proof : This follows from the fact that T_1 and T_2 are isomorphisms from $H^{1/2}(S)$ onto $H^{-1/2}(S)$ and that T_1 and $-T_2$ are coercive on $H^{1/2}(S)$ (signs depend on the orientation of ν). It follows that the mapping $f\mapsto\varphi(\varepsilon)$ which is linear and continuous from $L^2(\Omega)$ in $H^{1/2}(S)$ has analytic dependance in ε for $\varepsilon>0$. Remembering (2) the proof is complete∎

Corrollary 5: For every $k\geq 1$ the function

$$\varepsilon\mapsto\lambda_k(\varepsilon)$$

is analytic in $\varepsilon>0$.

Remark 6: This method allows one to prove analyticity up to $\varepsilon=0$ but akwardly only in the variable $\sqrt{\varepsilon}$ and this is why we shall have to work in a non self-adjoint setting to derive actual analyticity in ε.

4. Nonselfadjoint setting: Now we are going to study the inverse of the unbounded operator A_ε defined by the variationnal problem (1). We are going to work in the fixed space $H_1\equiv L^2(\Omega)$. This means that we are willing to consider **A_ε as nonselfadjoint**. However we do not lose the basic fact that its spectrum is **semi-simple**.

For every $f \in L^2(\Omega)$ there exists a unique $u(\varepsilon) \in V$ solution of

$$(5) \qquad \forall v\in V: \ a_\varepsilon(u(\varepsilon),v) = b_\varepsilon(f,v).$$

The inverse of the operator A_ε is

$$\Sigma_\varepsilon : f \mapsto u(\varepsilon),$$

where $u(\varepsilon)$ is solution of

$$u(\varepsilon)\in H^1{}_0(\Omega),\ -\mathrm{div}(\sigma_\varepsilon \nabla u(\varepsilon))=\rho_\varepsilon f.$$

We now proceed as in the preceding §, with the same notation for ν, G_i, P_i, T_i and g_i. Denoting $u_i(\varepsilon)$ the restriction of $u(\varepsilon)$ to Ω_i, this is equivalent to the transmission problem:

$u_i(\varepsilon) \in H^1(\Omega_i)$, $i=1,2$,

$u_i(\varepsilon) = 0$ on Γ_i, $i=1,2$,

$-\Delta u_1(\varepsilon) = f_1$ on Ω_1,

$-\varepsilon^{m-1}\Delta u_2(\varepsilon) = f_2$ on Ω_2,

$\varepsilon^m(\partial u_2(\varepsilon)/\partial\nu)=\partial u_1(\varepsilon)/\partial\nu$ on S.

Now consider

$$\psi(\varepsilon) = u_1(\varepsilon)|_S = u_2(\varepsilon)|_S$$

as **auxiliary unknown** one gets

$$(6)\qquad u_1(\varepsilon) = G_1f_1 + P_1\psi(\varepsilon),\ u_2(\varepsilon) = \varepsilon^{1-m}G_2f_2 + P_2\psi(\varepsilon).$$

and

$$(7)\qquad \varepsilon^m(\partial/\partial\nu)\{\varepsilon^{1-m}G_2f_2 + P_2\psi(\varepsilon)\} = (\partial/\partial\nu)\{G_1f_1 + P_1\psi(\varepsilon)\} \text{ on } S.$$

Equation (7) is equivalent to

$$(8)\quad (T_1-\varepsilon^mT_2)\psi(\varepsilon) = -g_1+\varepsilon g_2.$$

We are now going to prove the

Proposition 6 : The function $\varepsilon \mapsto \varepsilon^{m-1}\Sigma_\varepsilon$ has an analytic continuation to a complex neighborhood of 0 with values in $\mathcal{L}(H_1)$.

Proof : This follows from the fact that T_1 is an isomorphism from $H^{1/2}(S)$ onto $H^{-1/2}(S)$. Equation (8) is equivalent to

$$(I-\varepsilon^mT_1{}^{-1}T_2)\psi(\varepsilon) = -T_1{}^{-1}g_1+\varepsilon T_1{}^{-1}g_2.$$

Therefore φ depends analytically on ε for

$$|\varepsilon| < \|T_1{}^{-1}T_2\|^{-1}$$

where the norm is the operator norm in $H^{1/2}(S)$. In addition one has

$$\psi(\varepsilon) = (I-\varepsilon^mT_1{}^{-1}T_2)^{-1}T_1{}^{-1}[-(\partial/\partial\nu)G_1f_1+\varepsilon(\partial/\partial\nu)G_2f_2].$$

Taking (6) into account the proof is complete∎

5. The non degenerate case where $m=1$. It is possible to calculate explicitly the operator Σ_0 defined by analytic continuation. When $\varepsilon=0$ (8) means

$$T_1\varphi(0)= -g_1.$$

The function $u_1(0)$ given by (6) is solution of

$$(\partial/\partial\nu)(G_1f_1+P_1\varphi(0)|_S = g_1+T_1\varphi(0) = 0.$$

In other words one has

$$u_1(0) \in H^1(\Omega_1)$$
$$-\Delta u_1(0) = f_1 \text{ in } \Omega_1,$$
$$u_1(0) = 0 \text{ on } \Gamma_1$$
$$(\partial/\partial\nu)u_1(0) = 0 \text{ on } S.$$

Then $u_2(0)$ given by (6) is solution of

$$u_2(0) \in H^1(\Omega_2)$$
$$-\Delta u_2(0) = f_2 \text{ in } \Omega_2,$$
$$u_2(0) = 0 \text{ on } \Gamma_2,$$
$$u_2(0) = u_1(0) \text{ on } S.$$

Now it is convenient to consider linear continuous operators in $L^2(\Omega)$ as operator matrices corresponding to the isomorphism

$$L^2(\Omega)\equiv L^2(\Omega_1)\times L^2(\Omega_2).$$

The matrix of Σ_0 is

$$\begin{matrix} K_1 & 0 \\ P_2\gamma K_1 & G_2 \end{matrix}$$

where γ denotes the trace operator on S and K_1 the Green operator of the Neumann problem on Ω_1. In other words $v = K_1f$ is solution of

$$v \in H^1(\Omega_1),$$
$$-\Delta v = f, \text{ in } \Omega_1,$$
$$v = 0 \text{ on } \Gamma_1,$$
$$\partial v/\partial\nu = 0 \text{ on } S.$$

The operator Σ_0 has a very simple structure. It is triangular with symmetric compact entries on the diagonal. Its spectrum is a sequence of eigenvalues with finite multiplicities that tend to 0. The corresponding eigenvalues define a basis of H_1 but the coupling term $P_2\gamma K_1$ prevent them from being orthogonal.

At this stage the Proof of Theorem 1 is easy : Σ_ε is an analytic perturbation of Σ_0. Under our assumptions $\mu_0 = 1/\lambda_0$ is a simple eigenvalue of Σ_0. Following Kato (1966) there exists an analytic branch $\varepsilon \mapsto \mu(\varepsilon)$ of simple eigenvalues of Σ_ε coming from μ_0. We conclude by setting $\lambda(\varepsilon) = 1/\mu(\varepsilon)$ and defining $w(\varepsilon)$ as the corresponding branch of unitary eigenvectors of Σ_ε ∎

6. The degenerate case where m=2. In this case $R_\varepsilon = \varepsilon\Sigma_\varepsilon$ has an analytic continuation to a neighborhood of 0. We can calculate explicitly the first two terms ot the Taylor expansion of R_ε at 0.

Let us first calculate R_0. From (8) with $\varepsilon=0$ one deduces again that

$$T_1\varphi(0) = -g_1.$$

Then from (6) one deduces that

$$\varepsilon u_1(\varepsilon) = \varepsilon G_1 f_1 + \varepsilon P_1\varphi(\varepsilon) \longrightarrow 0$$

$$\varepsilon u_2(\varepsilon) = G_2 f_2 + \varepsilon P_2\varphi(\varepsilon) \longrightarrow G_2 f_2$$

as $\varepsilon \longrightarrow 0$ and consequently the matrix of R_0 is

$$\begin{matrix} 0 & 0 \\ 0 & G_2 \end{matrix}$$

In addition one has

$$\{\varepsilon u_1(\varepsilon) - 0\}/\varepsilon = u_1(\varepsilon) = G_1 f_1 + P_1\varphi(\varepsilon) \longrightarrow v_1 = G_1 f_1 + P_1\varphi(0)$$

$$\{\varepsilon u_2 - G_2 f_2\}/\varepsilon = P_2\varphi(\varepsilon) \longrightarrow v_2 = P_2\varphi(0)$$

as $\varepsilon \longrightarrow 0$. Since

$$(\partial/\partial\nu)(G_1 f_1 + P_1\varphi(0)|_S = g_1 + T_1\varphi(0) = 0,$$

the function v_1 is solution of

$$v_1 \in H^1(\Omega_1),$$

$$-\Delta v_1 = f_1, \text{ in } \Omega_1,$$

$$v_1 = 0 \text{ on } \Gamma_1,$$

$$\partial v_1/\partial\nu = 0 \text{ on } S$$

while v_2 is solution of

$$v_2 \in H^1(\Omega_2),$$
$$-\Delta v_2 = 0, \text{ in } \Omega_2,$$
$$v_2 = 0 \text{ on } \Gamma_2,$$
$$v_2 = v_1 \text{ on } S$$

In conclusion the matrix of R_ε is

$$\begin{matrix} \varepsilon K_1 & 0 \\ \varepsilon \Gamma_2 \gamma K_1 & G_2 \end{matrix}$$

up to $\varepsilon^2 U(\varepsilon)$ where $U(\varepsilon)$ is an analytic function in a neighborhood of 0, with values in $\mathcal{L}(H_1)$.

It is easy to prove Theorem 2 as in §5 : Under our assumptions $\mu_0 = 1/\lambda_0$ is a simple eigenvalue of G_2. Since μ_0 is strictly positive, μ_0 is also a simple eigenvalue of R_0. Following Kato (1966) there exists an analytic branch $\varepsilon \mapsto \mu(\varepsilon)$ of simple eigenvalues of $R_\varepsilon = \varepsilon S_\varepsilon$ coming from μ_0. It follows that $\mu(\varepsilon)/\varepsilon$ is a simple eigenvalue of S_ε and $\lambda(\varepsilon) = \varepsilon/\mu(\varepsilon)$ is a simple eigenvalue of A_ε i.e. eigenvalue of problem (1). Clearly $\lambda(0)=0$ and $\lambda(\varepsilon)$ depends analytically of ε in a neighborhood of 0. Finally one has $\lambda(\varepsilon)/\varepsilon \to 1/\mu_0 = \lambda_0$ as $\varepsilon \to 0$. We define $w(\varepsilon)$ as the corresponding branch of unitary eigenvectors of R_ε in H_1 ∎

Let us now prove Theorem 3 : The advantage of the 1d case is that the eigenvalues are simple.

Let us denote by $\lambda_1(\varepsilon) < \lambda_2(\varepsilon) < \ldots < \lambda_k(\varepsilon) \leqslant \ldots$ the eigenvalues whose existence has been established in Theorem 2. One has

$$\lambda_k(\varepsilon) = \varepsilon \xi_k + O(\varepsilon^2)$$

where $\xi_k \in \sigma(-\Delta_D, \Omega_2)$ and for a fixed ε one has $\lambda_k(\varepsilon) \to +\infty$ as $k \to +\infty$.

Now consider another eigenvalue $\lambda(\varepsilon)$ (by Corrolary 5 it is continuous in $\varepsilon > 0$) and assume that it has a limit λ_0 in $\varepsilon = 0$.

By continuity ans by the simplicity of the eigenvalues there exists a k such that

$$\lambda(\varepsilon) < \lambda_k(\varepsilon).$$

It follows that $\lambda(\varepsilon)/\varepsilon$ remains bounded as $\varepsilon \to 0$. Consequently λ_0 is 0.

Let $w(\varepsilon)$ be a unitary eigenvector corresponding to $\lambda(\varepsilon)$. One can select a sequence of values of ε tending to 0 and such that

$$\lambda(\varepsilon)/\varepsilon \to \eta \geqslant 0$$

$$w(\varepsilon) \to w$$

weakly in $H_1 = L^2(\Omega)$.

From the explicit form of Σ_ε one deduces that

$$\varepsilon K_1 w_1(\varepsilon) = (\varepsilon/\lambda(\varepsilon)) w_1(\varepsilon) + O(\varepsilon^2)$$

$$\varepsilon P_2 \gamma K_1 w_1(\varepsilon) + G_2 w_2(\varepsilon) = (\varepsilon/\lambda(\varepsilon)) w_2(\varepsilon) + O(\varepsilon^2).$$

This implies the strong convergence :

$$w_1(\varepsilon) = \lambda(\varepsilon) K_1 w_1(\varepsilon) + \lambda(\varepsilon) O(\varepsilon) \to 0$$

since K_1 is compact.

Then one has

$$\lambda(\varepsilon) P_2 \gamma K_1 w_1(\varepsilon) + (\lambda(\varepsilon)/\varepsilon) G_2 w_2(\varepsilon) = w_2(\varepsilon) + \lambda(\varepsilon) O(\varepsilon).$$

This shows the strong convergence :

$$w_2(\varepsilon) \to w_2(0) = \eta G_2 w_2(0)$$

since G_2 is compact.

One has necessarily $\|w_2(0)\|_{L^2(\Omega_2)} \neq 0$ since $w_1(0) = 0$ and $\|w(0)\|_{L^2(\Omega_2)} = 1$. This implies that $1/\eta$ is eigenvalue of G_2 and therefore

$$\eta \in \sigma(-\Delta_D, \Omega_2).$$

Thus we have found again the eigenvalues whose existence had already been proved in Theorem 2∎

II. 1d–2d Junctions.

<u>1.Introduction and notation</u> : Here, from a strictly mathematical point of view, we treat a stiff eigenvalue problem for a body involving a junction between components of different dimensions. There is no attempt to give a precise mechanical meaning to the problem under consideration. This problem was considered in <u>Lions</u> (1988). The behavior of the eigenvalues as $\varepsilon \longrightarrow 0$ is quite similar to what has been derived in part I.

Therefore we consider a pair of open sets of different dimensions : $\Omega_1 =]0,1[$ and $\Omega_2 \subset \mathbb{R}^2$ and fix $g \in H^{-1}(\Omega_2)$. The functions v under consideration are defined on $\Omega = \Omega_1 \cup \Omega_2$. They have two "components" $u_1 : \Omega_1 \mapsto \mathbb{R}$ and $u_2 : \Omega_2 \mapsto \mathbb{R}$.

Let ε be a strictly positive real number and let m be either 1 or 2.

We are going to consider the following variationnal problem in the space $H_\varepsilon = L^2(\Omega)$ equipped with the scalar product

$$b_\varepsilon(u,v) = \int_\Omega \rho_\varepsilon uv\, dx = \varepsilon \int_0^1 u_1 v_1\, dx + \int_{\Omega_2} u_2 v_2\, dx$$

$$\text{where } \rho_\varepsilon(x) = \varepsilon \text{ for } x \in \Omega_1 \text{ and } \rho_\varepsilon(x) = 1 \text{ for } x \in \Omega_2$$

The energy-space V is

$$V = \{v;\ v_1 \in H^1(\Omega_1), v_1(1) = 0, v_2 \in H^1{}_0(\Omega_2), v_1(0) = \langle v_2; g\rangle\}$$

while the energy-form is

$$a_\varepsilon(u,v) = \int_\Omega \sigma_\varepsilon \nabla u . \nabla v\, dx = \varepsilon^m \int_0^1 u_1' v_1'\, dx + \int_{\Omega_2} \nabla u_2 . \nabla v_2\, dx$$

$$\text{where } \sigma_\varepsilon(x) = \varepsilon^m \text{ for } x \in \Omega_1 \text{ and } \sigma_\varepsilon(x) = 1 \text{ for } x \in \Omega_2.$$

2. Scope : We wish to study the eigenelements of this problem **in dependance of** ε **tending to** 0, i.e. the $\lambda \in \mathbb{R}$ and the $w \in V$, $w \neq 0$, such that

(1) $$\forall v \in V: \quad a_\varepsilon(w,v) = \lambda b_\varepsilon(w,v).$$

3. Preliminaries : We consider the inverse of the unbounbded operator A_ε defined by the variationnal problem (1) in the **fixed** Hilbert space $H_1 \equiv L^2(\Omega)$.

For every $f \in L^2(\Omega)$ there exists a unique $u(\varepsilon) \in V$ solution of

(2) $$\forall v \in V: \quad a_\varepsilon(u(\varepsilon), v) = b_\varepsilon(f, v).$$

The inverse of A_ε is

$$S_\varepsilon : f \mapsto u(\varepsilon),$$

where $u(\varepsilon) \in V$ is solution of

$$\varepsilon^m \int_0^1 u_1(\varepsilon)' v_1' \, dx + \int_{\Omega_2} \nabla u_2(\varepsilon) . \nabla v_2 \, dx = \varepsilon \int_0^1 f_1 v_1 \, dx + \int_{\Omega_2} f_2 v_2 \, dx$$

for every $v \in V$.

Among the test functions there are the following particular ones

$$v = \{ v_1, 0 \}$$

with $v_1 \in H^1{}_0(\Omega_1)$. Consequently one has

$$\varepsilon^m \int_0^1 u_1(\varepsilon)' v_1' \, dx = \varepsilon \int_0^1 f_1 v_1 \, dx$$

for every $v_1 \in H^1{}_0(\Omega_1)$. It follows that

$$-u_1(\varepsilon)'' = \varepsilon^{1-m} f_1$$

in Ω_1.

Among the test functions there are also the following particular ones

$$v = \{ \langle v_2 ; g \rangle (1-x) ; v_2 \}$$

with $v_2 \in H^1{}_0(\Omega_2)$. Consequently one has

$$-\varepsilon^m \int_0^1 u_1(\varepsilon)' \langle v_2 ; g \rangle \, dx + \int_{\Omega_2} \nabla u_2(\varepsilon) . \nabla v_2 \, dx = \varepsilon \int_0^1 \langle v_2 ; g \rangle (1-x) \, f_1 \, dx + \int_{\Omega_2} f_2 v_2 \, dx$$

for every $v_2 \in H^1{}_0(\Omega_2)$. Since

$$\int_0^1 u_1(\varepsilon)' \, dx = u_1(\varepsilon)(1) - u_1(\varepsilon)(0) = -\langle u_2 ; g \rangle;$$

we have

$$\varepsilon^m \langle u_2(\varepsilon), g \rangle \int_{\Omega_2} g v_2 \, dx + \int_{\Omega_2} \nabla u_2(\varepsilon) . \nabla v_2 \, dx = \varepsilon \int_0^1 (1-x) \, f_1 \, dx \int_{\Omega_2} g v_2 \, dx + \int_{\Omega_2} f_2 v_2 \, dx$$

for every $v_2 \in H^1{}_0(\Omega_2)$ hence

$$u_2 \in H^1{}_0(\Omega_2)$$

$$-\Delta u_2(\varepsilon) + \varepsilon^m \langle u_2(\varepsilon) ; g \rangle g = f_2 + \varepsilon \{ \int_0^1 (1-x) \, f_1 \, dx \} g.$$

THis last equation has the form

(3) $$B u_2(\varepsilon) + \varepsilon^m C u_2(\varepsilon) = F(\varepsilon)$$

where B is an isomorphisme and C a linear continuous operator from $X := H^1{}_0(\Omega_2)$ into X' while $\varepsilon \mapsto F(\varepsilon)$ is an analytic function with values in X', for fixed $f \in L^2(\Omega)$.

We are now going to prove

Proposition 1 : The function $\varepsilon \mapsto \varepsilon^{m-1}S_\varepsilon$ has an analytic continuation to a complex neighborhood of 0 with values in $\mathcal{L}(H_1)$.

Proof : Equation (3) is equivalent to

$$u_2(\varepsilon) = (I+\varepsilon^m B^{-1}C)^{-1}B^{-1}F(\varepsilon)$$

and consequently $\varepsilon \mapsto u_2(\varepsilon)$ is analytic (with values in X and consequently also in $L^2(\Omega_2)$).

Then we write that $u_1(\varepsilon) \in H^1(\Omega_1)$ fulfils

(4) $$-u_1(\varepsilon)'' = \varepsilon^{1-m}f_1,\ u_1(\varepsilon)(0) = \langle u_2(\varepsilon);g\rangle,\ u_1(\varepsilon)(1) = 0.$$

It follows that $\varepsilon \mapsto \varepsilon^{1-m}u_1(\varepsilon)$ is analytic with values in $L^2(\Omega_1)$ ∎

4. The non degenerate case where $m=1$. The operator S_0 defined by analytic can be made explicit. When $\varepsilon=0$ (3) reads

$$Bu_2(0) = f_2.$$

Therefore

$$u_2 \in H^1{}_0(\Omega_2), -\Delta u_2 = f_2.$$

Then (4) reads

$$-u_1(0)'' = f_1,\ u_1(0)(0) = \langle u_2(0);g\rangle,\ u_1(0)(1) = 0.$$

In matrix forme S_0 is

$$\begin{matrix} G_1 & P_1G_2 \\ 0 & G_2 \end{matrix}$$

where G_i denotes the Green operator ot the Dirichlet problem in Ω_i i.e. $v = G_i f \in H^1{}_0(\Omega_i)$ and is solution of $-v''=f$ in Ω_1 if $i=1$ and $-\Delta v = f$ in Ω_2 if $i=2$, while P_1 denotes the operator

$$v \mapsto \langle v;g\rangle(1-x)$$

of $H^1{}_0(\Omega_2)$ into $L^2(\Omega_1)$.

Now it is easy to deduce the following Theorem where

$$\sigma(-\Delta_D;\Omega_i)$$

denotes the spectrum of the operator $-(\partial^2/\partial x^2)$ in Ω_1 if $i=1$ and of the operator $-\Delta$ in Ω_2 if $i=2$, both with a homogeneous Dirichlet condition.

S_0 is the the same as in §4 of the firs part. It follows in the same way that:

Theorem 2 : *We assume*

$$m=1,$$

$$\lambda_0 \in \sigma(-\Delta_D;\Omega_1) \cup \sigma(-\Delta_D;\Omega_2),$$

$$\lambda_0 \notin \sigma(-\Delta_D;\Omega_1) \cap \sigma(-\Delta_D;\Omega_2).$$

We assume in addition that λ_0 *is simple*. *Then problem* (1) *has an eigenvalue* $\lambda(\varepsilon)$ *and a coreresponding unitary eigenvector* $w(\varepsilon)$ *in* H_1 *which depends analytically on* ε *in a neighborhood of* $\varepsilon=0$ *and such that*

$$\lambda(\varepsilon) \longrightarrow \lambda_0$$

as $\varepsilon \longrightarrow 0$.

5. **The degenerate case** where $m=2$. In this case $R_\varepsilon = \varepsilon S_\varepsilon$ has an analytic continuation in a neighborhood of 0. The first two terms of the Taylor expansion of R_ε at 0 are

$$\begin{matrix} G_1 & \varepsilon P_1 G_2 \\ 0 & \varepsilon G_2 \end{matrix}$$

This S_ε is quite similar to the one found in §5 of the first part. In the sams way we prove the

Theorem 3 : *On suppose*

$$m=2,$$

$$\lambda_0 \in \sigma(-\Delta_D;\Omega_1)$$

We assume in addition that λ_0 *is simple*. *Then problem* (1) *has an eigenvalue* $\lambda(\varepsilon)$ *and a corresponding unitary eigenvector* $w(\varepsilon)$ *in* H_1 *which depends analytically on* ε *in a neighborhood of* $\varepsilon=0$ *and such that*

$$\lambda(\varepsilon)/\varepsilon \longrightarrow \lambda_0$$

when $\varepsilon \longrightarrow 0$.

III.References

Kato : Perturbation Theory for linear operators, Grundlehren, 132, Springer Verlag, 1966.

Lions : Cours au Collège de France, Paris, Octobre 1988.

Sanchez-Palencia : Non-homogeneous media and vibration theory, Lecture Notes in Physics, n°127, Springer Verlag, 1980.

Sanchez-Hubert and Sanchez-Palencia : Vibrations and coupling of continuous systems, Springer Verlag, 1989.

Département de Mathématiques, IMSP, Parc Valrose, 06034 NICE Cedex

V KOMORNIK

On the boundary stabilization of plates

Abstract. In [2], [3] a new direct approach was introduced for the boundary stabilization of the wave equation which, by avoiding the use of Liapunov functions, led to strong estimates of the decay rate of the energy. The aim of this note is to adapt this method for a simple plate model.

1. Introduction

Let Ω be a bounded domain in $\mathbf{R}^2$ with a boundary Γ of class C^4; its outward unit normal vector will be denoted by $\nu = (\nu_1, \nu_2)$. We shall also use the tangential unit vector $\tau = (-\nu_2, \nu_1)$. Let us assume that Ω is star-shaped with respect to some point $x^0 \in \mathbf{R}^2$ in the sense that putting

$$m(x) = x - x^0, x \in \mathbf{R}^2 \tag{1.1}$$

we have

$$m \cdot \nu \geq 0 \quad on \quad \Gamma \tag{1.2}$$

(the dot denotes the scalar product of $\mathbf{R}^2$).

Fix a positive number $\mu < 1$, three positive numbers p, q, r and consider the following system :

$$u'' + \Delta^2 u = 0 \quad in \quad \Omega \times (0, \infty) \tag{1.3}$$

$$\Delta u + (1-\mu) B_1 u = -p(m \cdot \nu) u_\nu \quad on \quad \Gamma \times (0, \infty) \tag{1.4}$$

$$(\Delta u)_\nu + (1-\mu)(B_2 u)_\tau = (m \cdot \nu)(qu' + ru) - p((m \cdot \nu) u_\tau)_\tau \quad on \quad \Gamma \times (0, \infty) \tag{1.5}$$

$$u(0) = u^0 \quad and \quad u'(0) = u^1 \quad on \quad \Omega; \tag{1.6}$$

here we use the notation

$$B_1 u = 2\nu_1\nu_2 \partial_1 \partial_2 u - \nu_1^2 \partial_2^2 u - \nu_2^2 \partial_1^2 u \tag{1.7}$$

$$B_2 u = (\nu_1^2 - \nu_2^2) \partial_1 \partial_2 u + \nu_1 \nu_2 (\partial_2^2 u - \partial_1^2 u) \tag{1.8}$$

and the subscripts ν and τ denote the normal and tangential derivations, respectively.

The system (1-3)–(1.6) represents a plate model studied earlier by Lagnese in [5].

We recall that for every $(u^0, u^1) \in H^2(\Omega) \times L^2(\Omega)$ the system (1.3)–(1.6) has a unique (weak) solution satisfying

$$(u, u') \in C([0, \infty); H^2(\Omega) \times L^2(\Omega)). \tag{1.9}$$

Furthermore, the system is dissipative in the following sense : defining the energy of the solutions by

$$E(t) = \frac{1}{2}\int_\Omega (u')^2 + Q(u)dx + \frac{1}{2}\int_\Gamma (m \cdot \nu)(p|\nabla u|^2 + ru^2)d\Gamma \tag{1.10}$$

where

$$Q(u) = (\partial_1^2 u)^2 + (\partial_2^2 u)^2 + 2\mu(\partial_1^2 u)(\partial_2^2 u) + 2(1-\mu)(\partial_1\partial_2 u)^2, \tag{1.11}$$

E(t) is a nonincreasing function in $[0, \infty)$. Indeed, this is an easy consequence of the following identity :

Lemma 1. For all $0 \leq S < T < \infty$ we have

$$E(S) - E(T) = q\int_S^T \int_\Gamma (m \cdot \nu)(u')^2 d\Gamma dt. \bullet \tag{1.12}$$

In fact, much stronger results hold if p and r are sufficiently small. Let us set

$$R = sup\{|m(x)| : x \in \Omega\} \tag{1.13}$$

and let a be the smallest positive constant satisfying

$$\int_\Omega (3u + m \cdot \nabla u)^2 dx \leq a^2 \Big[\int_\Omega Q(u)dx + \int_\Gamma (m \cdot \nu)(p|\nabla u|^2 + ru^2)d\Gamma\Big] \tag{1.14}$$

for all $u \in H^2(\Omega)$.

Theorem. Assume that

$$pR^2 \leq 2min\{\mu, 1-\mu\} \tag{1.15}$$
$$rR^2 \leq 2p \tag{1.16}$$

and set

$$T_0 = max\{\frac{1}{q} + \frac{R^2 q}{p}, 2a, \frac{3q}{r}\}.$$

Then for every $(u^0, u^1) \in H^2(\Omega) \times L^2(\Omega)$ the solution of (1.3)–(1.6) satisfies the estimates

$$E(t) \leq E(0)\exp(1 - \frac{t}{T_0}), for\ all\ t \in [0, \infty). \bullet \tag{1.18}$$

This theorem improves an earlier result of Lagnese [5], by weakening the conditions on p, r and by providing a smaller value for T_0. However, the purpose of this note is not to improve (sligthly) his result, but to give a new, simpler proof, which may be adapted for other problems as well.

In [4] a simple Liapunov type method was introduced for the study of the stabilization of the wave equation. This method was adapted and systematically used in [5] for the stabilization of various plate models. Using a further idea of Zuazua [11], [12] to apply more general feedbacks than in [4], it was shown in [2], [3] that, at least in the case of the wave equation, the use of Liapunov functions may be replaced by direct estimates, and that this modification leads to simpler proofs and at the same time to more precise (sometimes optimal) estimates. The purpose of this note is to adapt this method for the plate model (1.3)–(1.6).

Let us note that recently several other stabilization results were obtained for different plate models, using different methods by Bartolomeo, Lasiecka, Ourada and Triggiani, cf. [1], [6], [7], [9].

The author is grateful to I. Lasiecka and to J.-L. Lions for fruitful discussions.

2. Proof of the theorem

Every weak solution may be approximated in

$$C([0,\infty);H^2(\Omega))\cap C^1[0,\infty);L^2(\Omega))$$

by a sequence of strong solutions belonging to

$$C([0,\infty);H^4(\Omega))\cap C^1[0,\infty);H^2(\Omega))\cap C^2[0,\infty);L^2(\Omega)).$$

Therefore it is sufficient to prove the estimates (1.18) for strong solutions. Then we have the following two identities, cf. [5] :

Lemma 2. For all $0\le S<T<\infty$ we have

$$4\int_S^T\int_\Omega (u')^2+Q(u)dxdt=\Big[\int_\Omega 2u'Mudx\Big]_T^S$$
$$+\int_S^T\int_\Gamma (m\cdot\nu)\Big[(u')^2-Q(u)-2(qu'+ru)Mu-2p\nabla u\cdot\nabla(Mu)\Big]d\Gamma dt,$$

where we use the notation

$$Mu:=m\cdot\nabla u.\bullet$$

Lemma 3. For all $0 \leq S < T < \infty$ we have

$$2\int_S^T \int_\Omega Q(u) - (u')^2 dx dt + 2\int_S^T \int_\Gamma (m \cdot \nu)\Big[p|\nabla u|^2 + ru^2\Big] d\Gamma dt$$
$$= \Big[\int_\Omega 4uu' dx + q\int_\Gamma (m \cdot \nu) u^2 d\Gamma\Big]_T^S. \bullet$$

Remark. The lemmas 1–3 are proved by multiplying the equation (1.3) respectively with u', Mu, u and then by integrating by parts in $\Omega \times (S,T)$.•

Now we proceed as follows : combining the last two identities we have

$$\int_S^T \int_\Omega (u')^2 + 7Q(u) dx dt + 3\int_S^T \int_\Gamma (m \cdot \nu)\Big[p|\nabla u|^2 + ru^2\Big] d\Gamma dt$$
$$= \Big[\int_\Omega 2u'(3u + Mu) dx + \frac{3q}{2}\int_\Gamma (m \cdot \nu) u^2 d\Gamma\Big]_T^S$$
$$+ \int_S^T \int_\Gamma (m \cdot \nu)\Big[(u')^2 - Q(u) - 2(qu' + ru)Mu - 2p\nabla u \cdot \nabla(Mu)\Big] d\Gamma dt;$$

using (1.10) hence we obtain our basic inequality :

$$2\int_S^T E(t) dt$$
$$\leq \Big[\int_\Omega 2u'(3u + Mu) dx + \frac{3q}{2}\int_\Gamma (m \cdot \nu) u^2 d\Gamma\Big]_T^S$$
$$+ \int_S^T \int_\Gamma (m \cdot \nu)\Big[(u')^2 - Q(u) - 2(qu' + ru)Mu$$
$$-2p\nabla u \cdot \nabla(Mu) - 2p|\nabla u|^2 - 2ru^2\Big] d\Gamma dt.$$

Now we use (1.15) and (1.16) to majorize the last integral in this identity. We have

$$-Q(u) - 2p\nabla u \cdot \nabla(Mu)$$
$$= -Q(u) - 2p|\nabla u|^2 - 2p(\nabla u) \cdot (m \cdot \nabla(\nabla u)) \leq -2p|\nabla u|^2$$

and

$$-2(qu' + ru)Mu - 2p|\nabla u|^2 - 2ru^2$$
$$\leq 2R|qu' + ru||\nabla u| - 2p|\nabla u|^2 - 2ru^2 \leq \frac{R^2}{2p}(qu' + ru)^2 - 2ru^2$$
$$\leq \frac{R^2}{p}(q^2(u')^2 + r^2u^2) - 2ru^2 \leq \frac{R^2q^2}{p}(u')^2.$$

Hence, using (1.2) and Lemma 1 the last term of our basic inequality is majorized by

$$(1+\frac{R^2q^2}{p})\int_S^T\int_\Gamma(m\cdot\nu)(u')^2d\Gamma dt=(\frac{1}{q}+\frac{R^2q}{p})(E(S)-E(T)).$$

Furthermore, using (1.14) we have

$$\int_\Omega 2u'(3u+Mu)dx\le 2aE$$

and therefore the first term on the right hand side of the basic inequality is majorized by

$$\max\{2a,\frac{3q}{r}\}(E(S)+E(T)).$$

Using these estimates and (1.17), from the basic inequality we conclude that

$$\int_S^T E(t)dt\le T_0E(S)\qquad for\qquad all\qquad 0\le S<T<\infty;$$

letting $T\longrightarrow\infty$ hence we obtain that

$$\int_S^\infty E(t)dt\le T_0E(S)\qquad for\qquad all\qquad S\in[0,\infty).\tag{2.1}$$

Now the estimates (1.18) follow easily from (2.1), as in [3]. Indeed, (2.1) implies that

$$\frac{d}{ds}\left[\exp(\frac{s}{T_0})\int_S^\infty E(t)dt\right]\le 0\qquad in\qquad (0,\infty)$$

and therefore

$$\int_S^\infty E(t)dt\le\exp(-\frac{s}{T_0})\int_0^\infty E(t)dt\qquad for\qquad all\qquad S\in[0,\infty).$$

Using (2.1) with S=0 we obtain that

$$\frac{1}{T_0}\int_S^\infty E(t)dt\le E(0)\exp(-\frac{s}{T_0})\qquad for\qquad all\qquad S\in[0,\infty).$$

Since E(t) is nonnegative and nonincreasing, the left hand side is minorized by

$$\frac{1}{T_0}\int_S^{S+T_0}E(t)dt\ge E(S+T_0)$$

whence, putting $t := S + T_0$, (1.18) follows for all $t \geq T_0$. Since (1.18) is obvious for $t < T_0$ ($E(t)$ is nonincreasing), the proof of the theorem is completed.•

References

[1] J. Bartolomeo and R. Triggiani, Uniform energy decay rates for Euler–Bernoulli equations with feedback operators in the Dirichlet/Neumann boundary conditions, to appear.

[2] V. Komornik, Stabilisation frontière rapide de l'équation des ondes, C. R. Acad. Sci. Paris Sér. I Math. 309(1989), 483-486.

[3] V. Komornik, Rapid boundary stabilization of the wave equation, SIAM J.Control Opt., janvier 1991.

[4] V. Komornik and E. Zuazua, A direct method for the boundary stabilization of the wave equation, J. Math. Pures Appl. 69(1990), 33-54.

[5] J. Lagnese, Boundary Stabilization of Thin Plates, SIAM Studies in Appl. Math., SIAM, Philadelphia, 1989.

[6] I. Lasiecka, Exponential decay rates for the solutions of Euler–Bernoulli equations with boundary dissipation occuring in the moments only, to appear.

[7] I. Lasiecka and R. Triggiani, Exact controllability and uniform stabilization of Kirchoff plates with boundary control only on $\Delta w|_\Sigma$ and homogeneous boundary displacement, to appear.

[8] J.-L. Lions, Exact controllability, stabilizability and perturbations for distributed systems, SIAM Rev. 30(1988), 1-68.

[9] N. Ourada and R. Triggiani, Uniform stabilization of the Euler–Bernoulli equation with feedback operator only in the Neumann boundary condition, to appear.

[10] D. L. Russell, Controllability and stabilizability theory for linear partial differential equations. Recent progress and open questions, SIAM Rev. 20(1978), 639-739.

[11] E. Zuazua, Some remarks on the boundary stabilizability of the wave equation, Control of Boundaries and Stabilization, ed. J. Simon, Lecture Notes in Control and Information Sci. Vol. 125, Springer–Verlag, Berlin, 1989, 251-266.

[12] E. Zuazua, Robustesse du feedback de stabilisation par contrôle frontière, C. R. Acad. Sci. Paris Sér. I Math. 307(1988), 587-591.

Eötvös Loránd University, Department of Analysis, H-1088 Budapest, Múzeum krt. 6-8, Hungary and Université de Bordeaux I, U.F.R. de Mathématiques et d'Informatique, 351, cours de la Libération, 33405 Talence Cédex France

Current address : UFR de Mathématiques et d'Informatique, 7, rue René Descartes, 67084 Strasbourg Cédex France

J P ZOLESIO

Eulerian Galerkin approximation for wave equation in moving domain

CNRS, Departement de Mathematiques
Parc Valrose ,06034 Nice cedex, France

We consider a non linear wave equation in a moving domain for which the approximation by Galerkin method is an open question.Assuming homogeneous Dirichlet boundary conditions on the lateral boundary Σ (which is time like) we derive the galerkin approximation $y_m(t,x)=\mathbf{h}_{m,i}(t)w_i(t,x)$ of the solution assuming that the family w_i is convected.We show that ,under smoothness assumptions on Σ , such family always exist.The differential system whose the vector $\mathbf{h}_m(t)$ is solution contains two additive terms (by comparison to the usual "cylindrical" galerkin method) involving the two first time derivatives of the functions w_i (t,.). These terms explicitely depend on the choice of the vector field V convecting the family w_i. Examples are given.

I. Introduction and Main Results

We are concerned with the wave equation in a moving domain.The wellposdness of this problem has been studied by several authors and by different technics.The specific situation where the domain has monotonic variations leads to dissipatives solution depending on the lateral boundary conditions.For homogeneous Dirichlet boundary conditions on the lateral boundary of the evolution domain the energy is decaying when the domain is monotoniquely increasing.For homogeneous Newman boundary conditions the energy is decaying with the geometrical domain.In these very simple situations a constructive proof of existence for the solution of the wave equation is obtain using a "non moving" Galerkin approximation enhanced with a penalty technic.We briefly describe now this approach which was called the "complementary domain approach" by **J.L.Lions** [L].The moving domain which is assumed to be bounded and smooth is always contained in a fixed and bounded domain D.Then,as far as the homogeneous Dirichlet boundary condition is concerned ,one consider a dense family $w_1,...,w_m,...$ of $H^1(D)$ and writting the usual galerkine approximation related to the wave equation in the cylinder [0,T]xD one add a penalty term on the integral terms over the complementary in D of the moving domain.This approach has been revisited in J.P.Zolésio[Z1] and extended to Newman boundary conditions,also in J.P.Zolésio [Z2] non monotonic situations are treated for mixed Newman-Dirichlet boundary condition.This method furnishes an existence result for smooth initial data (y_0,y_1) in $H_0^1(\Omega_0)xL^2(\Omega_0)$ but as it obtains the solutions y as a weak-* limit it does not furnish the classical regularity $C^\circ([0,T],H^1(\Omega_t)) \cap C^1([0T],L^2(\Omega_t))$. The non monotonical situations have been studied by **Cooper and Straus** [CS],**Cooper and Bardos**[CB] using a change of variable f defined from the non cylindrical evolution domain Q which is the union of the $\{t\}x\Omega_t$ onto the cylinder $Q_0=[0,T]x\Omega_0$,they assume f to have the so called "hyperbolic" property which roughly speaking says that under the transportation by f the wave operator remains a wave operator,under this assumption they use the complementatry domain or Galerkin-penalty technic in Q_0 which furnishes a solution,without classical smoothness as we mentioned and then without any propagator for the solution.In **C.Bardos,G.Chen**[BC] the Homogeneous Dirichlet boundary condition is considered with an increasing domain,to obtain a solution they use the results of [CB] and they say that the classical smoothness $C^\circ([o,T],H^1(\Omega_t)) \cap C^1([o,T],L^2(\Omega_t))$ can be recover using the **J.Lions-.Magenes**[LM] lemma,most probabely on the transported solution on Q_0 (for in the moving domain framework the generalisation of this lemma is not obvious) and it seems that even in that situation the obtention of the propagator (or evolution operator U(t,s)) should be clarified .A completely different approach has been developed by **G.Da Prato and J.P.Zolésio** [DPZ1] where using the speed method and the Kato theory of non autonomeous semi groups well posness,unicity ,smoothness and propagator are obtained without any other hypothesis then the smoothness of the lateral boundary Σ (which is the union of the $\{t\}x\partial\Omega_t$) . See also [DPZ2],[DPZ3].At this point we recall an other technic succesfully applied to the damped wave equation and to the Riccati equation in a moving domain by **P.Cannarsa,G.Da Prato,J.P.Zolésio** [CDZ1],[CDZ2],[CDZ3],[CDZ4],[CDZ5] .this technic consists,as in the complementary domain approach,to consider the larger and fixed domain D and using the operator A(t) in each domain Ω_t extend A(t) in Å(t) whose domain $\mathfrak{D}(t)$ is moving in the fixed Hilbert space $H^1(D)xL^2(D)$,that is to change the moving domain Ωt in the fixed domain D for a moving domain $\mathfrak{D}(t)$ in the fixed space $H^1(D)xL^2(D)$.

These different results are not constructive for the solution y of the wave equation in the moving domain Ω_t .The Galerkin-penalty technic is not realistic in many situations and is not efficient in situations where the domain Ω_t is not monotoniquely varying.On the other hand the "hyperbolic" change of variable method is absolutely not constructive for there is no constructive results for the hyperbolic mapping f from Q onto Q_0;in fact in [CB] it was dicussed the fact that the existence of f is was an open question and that the difficulty was to be able to connect that problem to the geometrical properties of the lateral boundary Σ of the non cylindrical evolution domain Q.In a recent paper **Sikorav** [S] has given existence results for the mapping f.

Nevertheless this result is not constructive and even if it was the change of variable technic would not be realistic for example in the very important field of numerical approximation of the solution y.

While the non cylindrical evolution problem is not a shape optimization problem in the sens of Boundary Variation and Boundary Control [Z5] it is strongly connected for the difficulties which arise from the fact that the geometrical domain is moving are the same ones.In particular the crucial difficulty about the construction and the properties of transformations T_t mapping the initial domain Ω_0 onto the moving domain Ω_t is completely understood so that in this paper we shall begin by giving in a very short way the construction of the hyperbolic mapping f^{-1} in the following form : $f^{-1}(t,x)= T_t(x)$. The key idea developped in **J.P.Zolésio**[Z3],[Z4] after 1975 ,was to consider the moving domain Ω_t as the volume occupied by a fluid (or any continum media) with Lagrangian coordinate X in Ω_0 while $x=T_t(X)$ are the Eulerian coordinate (or actual coordinate in Ω_t).The parameter which caracterises the transformation is now the Speed Vector field $V(t,x) = (\partial_t T_t)o(T_t)^{-1}(x)$ .The transformation T_t can be recovered from the non autonomeous field :Tt is the flow of V,that is that Tt(X),where x=x(t,X) is the solution of the ordinary differential equation $\partial_t x=V(t,x),\ x(o)=X$, or equivalentely $x(t)=X+\int^t_0 V(s,x(s))ds$.

The aim of this paper is to built convergent Galerkin Methods in Eulerian coordinates.In some sens we could say that the complementary domain technic,or Galerkin-penalty technic,is a Galerkin technic in the lagrangian coordinates as the family $w_1,...w_m,...$ on which the problem will be projected could be any extension to D of a such family defined on the domain Ω_0.But now,as it is pratiquely done in many software (without any justification) ,we shall consider at each time t a family $w_1(t,.),...,w_m(t,.),...$ of functions defined on the domain Ω_t.In some example these function could be thought as beeing the m fist eigen functions of the Laplace operator in the domain Ω_t with the homogeneous Dirichlet conditions.We shall show in this paper that under the single condition that the lateral boundary Σ of the non cylindrical evolution operator is smooth enough and time like (that point will be clarified) the Galerkin approximation associated to the family $w_i(t,.)$ converges to the solution y of the wave problem in Q provided that the basis is a CONVECTED basis .This property eliminates the possibility of chosing the eigen functions,as it is in fact done in some software!Also we shall properly define what is the Galerkin approximation in Eulerian coordinate,associated to the convected moving basis $w_i(t,.)$.It turns out that the differential system that we shall write is not the usual one in which one would simply add the letter t in the w_i and the doain Ω;two new terms are introduced involving the first and second time derivativesof the moving basis $w_i(t,.)$

To make clear the modification introduced in the galerkine algorithm we shall begin by recalling
what is the usual galerkin method in the cylindrical situation,that is when the domain does not move with time.

I .1.The cylindrical Galerkin Method.

Ω_0 denotes a smooth bounded domain in R^n and given a final time T we set

$$Q_o=]0,T[x\Omega_o \quad , \Sigma_o =]0,T[\ x\partial\,\Omega_o \tag{1.1}$$

given initial data y_o in $H^1{}_o(\Omega)$and y_l in $L^2(\Omega_o)$,we consider the following wave problem

$$Hy\ (t,x) = \partial^2{}_{tt}\, y(t,x) -\Delta y(t,x)+ \ ß\ (|\ y\ |^p y)(t,x)= 0, \text{ in } \ Q_o \tag{1.2}$$

$$y(t,x)= 0 \text{ on } \Sigma_o \tag{1.3}$$

$$y(0,x)= y_0(x) \quad \text{on } \Omega_o \tag{1.4}$$

$$\partial_t y(0,x)=y_1(x) \quad \text{on } \Omega_o \tag{1.5}$$

Given a dense family $w_1,...,w_m,...$ dense in $H^1{}_o(\Omega)$
we consider the expension

$$y_m(t,x) = h_1(t)w_1(x)+...+h_m(t)w_m(x) \tag{1.6}$$

where the m real functions h_i are solution of the following ordinary differential system :

$$\int_{\Omega o} H\, y_m(,x)\, w_i(x)\, dx =0 \qquad i=1,...,m \tag{1.7}$$

with the initial conditions

$$yi(0,x)=ai\ , \ \ \partial_t yi(0,x) = bi \tag{1.8}$$

with $a_1\, w_1+...+a_m w_m$ which converges in $H^1{}_0(\Omega)$,as m goes to infinity,to y_0
and $b_1\, w_1+...+b_m w_m$ converges in $L^2(\Omega)$ to y_1

The convergence of y_m to the unique solution y to problem (1.2)-(1.5) is classical. The system (1.7) can be equivalentely written as follows:

$$\partial_{tt}h_{m,j}(t) \int_\Omega w_j\, w_i\, dx + h_{m,j}(t) \int_\Omega (\partial_x w_j\, \partial_x w_i\, dx + ß\, |y_m|^p\, w_i)dx = 0 \qquad (1.9)$$

for all i,i=1,...,m

which can also be written

$$\partial_{tt}h(t) + M^{-1}(A + ß\ B(y_m)).h(t) = 0 \qquad (1.10)$$

where M is the usual mass matrix and A the rigidity matrix

I.2 The non cylindrical case and main result.

We consider the non cylindrical evolution domain

$$Q = \bigcup_{o<t<T} \{t\}x\,\Omega_t \qquad (1.11)$$

with the lateral boundary

$$\Sigma = \bigcup_{o<t<T} \{t\}x\partial\,\Omega_t \qquad (1.12)$$

we assume:

H1 Σ is a C^1 manifold (with boundary $\partial\Sigma = \{0\}x\Omega_o \cup \{T\}x\Omega_T$) in $\mathbf{R}^{n+1}$

H2 Σ is time like : if π denotes the unitary normal field on Σ out going to Q, π belonging to $C^o(\,cl(\Sigma),\mathbf{R}^{n+1})$,

$$\pi(t,x) = (1+v(t,x)^2)^{-1/2}\,(-v(t,x),\, n_t(x)\,) \qquad (1.13)$$

then v verifies:

$$|v(t,x)| < 1 \quad \text{on } cl(\Sigma) \qquad (1.14)$$

where $n_t(x)$ denotes the unitary normal field to $\partial\Omega_t$ out going to Ω_t, $n_t(.)$ belonging to $C^0(\,\partial\Omega_t,\mathbf{R}^n\,)$

H3 for each t , $w_1(t,.),...,w_m(t,.),...$is a dense family in $H^1{}_0(\Omega_t)$ which is convected: where exists a field V, element of $C^o([0,T],C^1(D,\mathbf{R}^n))$,where D is a compact set in

$\mathbf{R}^n$, intD being a neighborhood of $cl(\bigcup_{0<t<T} \Omega t)$, verifying

$$w_i(t,x)= w_i(o, T_t(V)^{-1}(x)) \quad \text{for all } (t,x) \text{ in } Q \text{ and } i=l,...,m,... \tag{1.15}$$

Where $T_t(V)$ is the flow associated to V ,T(V):t ------->$T_t(V)$ belongs to $C^{l(}([0,T], C^l(D,\mathbf{R}^n))$.

We also assume that $T_t(V)$ maps Ω_0 onto Ω_t , ∂D onto ∂D and D onto D that is that for all t,0<t<=T, we have $\Omega_t=T_t(V)(\Omega_o)$, $\partial\Omega_t=T_t(V)(\partial\Omega_o)$ (1.16)

H4 furthermore we assume that the field V is "hyperbolic":

$|V(t,x).n_t(x)|<l$ for all (t,x) on Σ implies $|V(t,x)|<l$ for all (t,x) in Q

We consider the non linear wave problem in the non cylindrical domain Q:find y in $E=L^\infty(0,T,H^1{}_0(\Omega_t))\ W^{1,\infty}(0,T,L^2(\Omega_t))$ solution of:

$$\mathbf{H}y\ (t,x) = \partial^2{}_{tt}\, y(t,x) - \Delta y(t,x) + \beta\,(|\,y\,|^p y)(t,x) = 0, \text{ in } Q \tag{1.17}$$

$$y(t,x)= 0 \text{ on } \Sigma \tag{1.18}$$

$$y(0,x)= y_0(x) \quad \text{on } \Omega_0 \tag{1.19}$$

$$\partial_t y(0,x)=y_1(x) \quad \text{on } \Omega_0 \tag{1.20}$$

Let

$$\mathbf{y}_m(t,x) = \mathbf{h}_{m,1}(t)w_1(t,x)+...+\mathbf{h}_{m,m}(t)w_m(t,x) \tag{1.21}$$

where $\mathbf{h}_m(t)=(\mathbf{h}_{m,1}(t),...,\mathbf{h}_{m,m}(t))$ is solution of the modified system:

$$\partial_{tt}\mathbf{h}_m(t) + M(t)^{-1}(A(t)+\mathbf{C}(t)+\beta\, B_t(y_m)).h(t) + 2M(t)^{-1}.\mathbf{D}(t).\partial_t\mathbf{h}_m(t)=0 \tag{1.22}$$

with

$$M(t)_{i,j}= \int_{\Omega_t} w_i(t,x)w_j(t,x)\, dx \tag{1.23}$$

$$A(t)_{i,j}= \int_{\Omega_t} \partial_x w_i(t,x)\partial_x w_j(t,x)\, dx$$

$$\mathbf{C}(t)_{i,j}= \int_{\Omega_t} \partial_{tt} w_i(t,x)\, w_j(t,x)\, dx \tag{1.24}$$

$$\mathbf{D}(t)_{i,j}= \int_{\Omega_t} \partial_t w_i(t,x)\, w_j(t,x)\, dx$$

and B_t is the non linear operator :

$$(p+1)\, B_t(z)_i = \int_{\Omega_t} |z(t,x)|^{p+1}\, w_i(t,x)\, dx \qquad (1.25)$$

Theorem 1. *Assume that (H1)-(H4) are satisfiyed, that* (y_0,y_1) *belongs to* $H^1{}_0(\Omega_0)xL^2(\Omega_0)$ *and that* β *is non negative.*
Then the problem (1.16)-(1.20) has a unique solution y in E ; the sequence y_m *defined at (1.21)-(1.25) converges to y in* $L^2(Q)$*.We have also the convergence weak-*$*$ *in* $L^{\approx}(0,T,L^2(\Omega_t))$ *of* $\partial_t y_m$ *to* $\partial_t y$ *and of* $\partial_x y_m$ *to* $\partial_x y$ *.*

Remark 1. The expression (1.13) for the unitary normal vector field π on Σ ,out going to Q,uniquely determined the term v on Σ ,v beeing an element of $C^o(\mathrm{cl}(\Sigma))$;the minus sign in (1.13) has been chosen such that the moving domain Ωt is monotoniquely varying when v does not change its sign :

Ω_t is increasing with t,$t_1<t<t_2$, iff v(t,x) is non negative for $t_1<t<t_2$ and x in $\partial\Omega_t$
Ω_t is decreasing with t,$t_1<t<t_2$, iff v(t,x) is non positive for $t_1<t<t_2$ and x in $\partial\Omega_t$
We say that v(t,.) is the normal componant of the speed of the boundary $\partial\Omega_t$,this terminology is justified by the following result:

Proposition 1 .
Let be given a field V, element of $C^o([0,T],C^1(D,\mathbf{R}^n))$,*where D is a compact set in* $\mathbf{R}^n$ *, intD being a neighborhood of* $\mathrm{cl}(\bigcup_{0<t<T} \Omega t)$*, the following two properties are equivalent :*

i) for all t ,0<t<T, $\Omega_t = T_t(V)(\Omega_0)$ and $\partial\Omega_t = T_t(V)(\partial\Omega_0)$ (1.26)
ii) for all t ,0<t<T, $V(t,x).n_t(x) = v(t,x)$. (1.27)

II. Explicit Expressions for the New Matrices C(t) and D(t)

Following assumptions (1.15) and (1.24) we can write the basis w_i in the following form:

$$w_i(t,x) = \mathbf{W}_i \circ T_t(V)^{-1} \qquad (2.1)$$

where

$$\mathbf{W}_i(x) = w_i(0,x) \qquad (2.2)$$

introducing the shift vector field V_t defined by:

$$V_t(s,x) = V(t+s,x) \qquad (2.3)$$

We have the propagator property:

$$T_{t+r}(V) = T_r(V_t)\circ T_t(V) \qquad (2.4)$$

From (2.1),(2.4) we get

$$w_i(t+r,x)=\mathbf{W}_i \circ T_t(V)^{-1} \circ T_r(V_t)^{-1}$$

that is

$$w_i(t+r,x)=w_i(t,x) \circ T_r(V_t)^{-1} \qquad (2.5)$$

and we have

$$\partial_t w_i(t,x)= -\partial_x w_i(t,x).V(t,x) \qquad (2.6)$$

$$\partial^2_{tt} w_i(t,x) = - < \partial_x w_i(t,x) , \partial_t V(t,x) >+ <\partial_x(\partial_x w_i(t,x).V(t,x)) , V(t,x) > \qquad (2.7)$$

and we get the following

Proposition 2. *Assuming that the speed vector field V belongs to* $C^1([0,T\},C^1(\mathbf{R}^n,\mathbf{R}^n))$ *let* $\partial_t V$ *denotes the* ***acceleration*** *field,*$\partial_t V(t,x)= (\partial/\partial t)V(t,x)$,*then we have*:

$$\mathbf{C}(t)_{i,j}= -\int_{\Omega_t} \partial_x w_i(t,x).\partial_t V(t,x) w_j(t,x)\, dx -\int_{\Omega_t} \partial_x w_i(t,x).V(t,x)\, div(V(t,x)w_j(t,x))\, dx \quad (2.8)$$

and

$$\mathbf{D}(t)_{i,j} =- \int_{\Omega_t} \partial_x w_i(t,x).V(t,x)w_j(t,x)\, dx \qquad (2.9)$$

Remark 2. The field V(t,.) and the acceleration term $\partial_t V(t,.)$ appear in the expressions of **C**(t) and **D**(t) only through theire restrictions to the domain Ω_t .

We can give examples of non autonomeous vector fields V verifying (1.26) and (1.27) : the simplest idea is to search V(t,.) as deriving in Ω_t from a scalar potential $\mathbf{P}_t$

$$V(t,x)|_{\Omega t}= \text{grad } \mathbf{P}_t(x) \quad \text{in } \Omega_t \qquad (2.10)$$

with the boundary condition:

$$\partial_{n_t} P_t = v(t,.) \text{ on } \partial\Omega_t \qquad (2.11)$$

where n_t is the unitary normal field on $\partial\Omega_t$ out going to Ω_t .To determine the potential $\mathbf{P}_t$ one can for example assume $\mathbf{P}_t$ to be solution in Ω_t of some elliptic problem

$$\cdot\Delta \mathbf{P}_t = f(t) \text{ in } \Omega_t \qquad (2.12)$$

where f(t) only depends on t and is chosen to verify the compatibility condition for the Neumann problem (2.11),(2.12) : $f(t)\,|\Omega_t| + \int_{\partial\Omega t} v(t,x)\, d\mu_t(x) = 0$

where $|\Omega_t|$ is the measure of Ω_t and μ_t is the surface measure on the boundary .When Ω_t has a fixed measure then f can be taken equal to zero that is to say that $\mathbf{P}_t$ is harmonic in Ω_t .

On the whole domain D the field V(t,.) will be defined as any smooth extention of $V(t,x)|_{\Omega t}$ to D . In that situation the acceleration term $\partial_t V(t,.)$ restricted to the domain Ω_t will be ,by the very definition [Z 4],the so-called "shape derivative" of grad **P** in the direction of the vector field.For convenience we shall recall some results :

We assume given a smooth bounded domain Ω , Ω included in D,and a smooth vector field V belonging to $C^o([o,t^*[,C^l(D,\mathbf{R}^n))$ vith v(t,.)=0 on ∂D .For each t in some neighborhood of zero is defined $\Omega_t = T_t(V)(\Omega)$ and we suppose given an element $\mathbf{P}_t$ of the Sobolev space $H^{s+l}(\Omega_t)$

Proposition 3 . *Assume s to be a non negative real number and*

(SDl) There exists an extention $\mathbb{P}(t,.)$ of $\mathbf{P}_t$,$\mathbb{P}(t,.)$ belonging to $H^{s+l}(D)$,

for any x in Ω_t, $\mathbb{P}(t,x) = \mathbf{P}_t(x)$

(SD2) $\mathbb{P}(t)$ belongs to $C^l([0,t^*[, H^s(D))$, where $\mathbb{P}(t)(x) = ,\mathbb{P}(t,x)$

Then $\partial_t\ \mathbb{P}(t)$ *belongs to* $C^0([0,t^*[, H^s(D))$ *and **the restriction*** $[\partial_t\ \mathbb{P}(t)]_{|\Omega_t}$ *is for each t an element of* $H^s(\Omega_t)$ *which is **independant** on the choice of the extension* $\mathbb{P}(t,.)$ of $\mathbf{P}_t$.

Definition l.

Under assumptions (SDl) and (SD2) the Shape Derivative of **P** in $H^s(\Omega)$, at Ω and in the direction of the vector field V, is given by

$$\mathbf{P}'(\Omega;V) = [\partial_t\ \mathbb{P}(t)]_{|\Omega,t=0}\ ,\quad \text{element of } H^s(\Omega)\quad . \qquad (2.13)$$

Remark 3.

From the property (2.3)-(2.4) the Shape Derivative of P_t in $H^s(\Omega_t)$,at Ω_t in the direction of the the shift vector field V_t is given by:

$$\mathbf{P}'(\Omega_t;V_t) = [\partial_t\ \mathbb{P}(t)]_{|\Omega}\ ,\quad \text{element of } H^s(\Omega_t)\quad . \qquad (2.14)$$

The acceleration term in Ω_t can be taken as the Shape Derivative ,for some suitable s,of grad $\mathbf{P}_t$.

For this purpose we have the following result.

Proposition 4.

Assume s-l *to be a non negative real number and* **P** *to be Shape Differentiable in* $H^s(\Omega)$, at Ω *in the direction of the field V. Then grad* ***P*** *is Shape Differentiable in* $H^{s-l}(\Omega)$, *at* Ω *in the direction of the fiels V and the Shape Derivative is given by:*

$$(\text{ grad } \mathbf{P})'(\Omega;V) = \text{grad}[\ \mathbf{P}'(\Omega;V)\ ]\ ,\ \text{element of } H^{s-l}(\Omega)\quad . \qquad (2.15)$$

Finely let us recall that the shape derivative **P'** is chartacterised as the solution of a Neuman problem in Ω_t and more precisely we have the following results.

Proposition 5.

We assume the lateral boundary Σ *to be smooth enough,and we consider an extension N(t,x) of the unitary normal field* $n_t(x)$ *on* $\partial\Omega_t$,*that is N belongs to* $C^l([0,T]xD,\mathbf{R}^n)$ *such that*

$N(t,x) = n_t(x)$ for all x in $\partial\Omega_t$

Also **v** *is a smooth extension of v(t,.),that is* **v** *belongs to* $C^1([0,T]xD)$ *such that*

$\mathbf{v}(t,x) = v(t,x)$ for all x in $\partial\Omega_t$

We consider the field Z= **v** *N defined on [0,T]xD ,which is admissible in the sens that it verifies the two equivalent conditions of Proposotion 1.The mapping* $r \dashrightarrow v(t+r,.)\circ T_r(Z_t)$ *belongs to* $C^1([0,t^*[,C°(\partial\Omega_t))$ *, its derivative at r= 0,* $v^{\cdot}(t,.) = [\,\partial_r v(t+r,T_r(Z_t)(.)\,)]_{r=0}$ *element of* $C°(\partial\Omega_t)$ ***does not depends*** *on the choice of the extensions N and* **v** *which define the field Z.*

Definition 2. This derivative $v^{\cdot}(t,.)$ is the boundary acceleration.

Proposition 6.

The solution **P** to the Neuman problem has a Shape Derivative **P'** (Ω;V) which is the solution of the following Neuman problem :

$$- \Delta \mathbf{P'} = f'(t) \text{ in } \Omega_t \qquad (2.16)$$

$$\partial_{n_t}\mathbf{P'} = -\mathrm{div}_{\partial\Omega_t}(\, v(t,.)\ \mathrm{grad}_{\partial\Omega_t}\mathbf{P}_t\,) + f(t)v(t,.) + H(t,.)v(t,.)^2 + v^{\cdot}(t,.) \text{ on } \partial\Omega_t \qquad (2.17)$$

where $\mathrm{grad}_{\partial\Omega_t}$ is the tangential componant of the gradient , $\mathrm{div}_{\partial\Omega_t}$ is the tangential divergence operator on the boundary $\partial\Omega_t$ and H(t,.) is the mean curvature on $\partial\Omega_t$.

Remark 4. From the knowlege of the normal boundary speed v(t,.) and acceleration $v^{\cdot}(t,.)$ the mappings f and f' are obtained and then we get **P'** as solution of (2.16)-(2.17) and the Speed and the acceleration as V= grad**P** and ∂_tV= grad **P'** in Ω_t so that the matrices **C**(t) and **D**(t) can be derived.We asumed in that example that $\|\,\mathrm{grad}\mathbf{P}_t\,\|_{L^\infty(\Omega)} \leq \mu < 1$,for all t .If this condition is violated one can chose P as the solution of the variational inequality (imposing that unilateral condition) but now the expression of the acceleration is more complicate,see **J.Sokolowski-J.P.Zolésio** [SZ]

III. Convergence of the Eulerian Galerkin Approximation .

In this section we provide the proof of theorem 1 which is the main result.It will be done in three steps.

3.1 The Transported Wave Equation to the Cylindrical Domain Q_0 .

Let V be an admissible field, smooth enough say in $C^1([0,T],C^1(D,\mathbf{R}^n))$ with V(t,.)=0 on ∂D , and such that its flow T_t maps Ω_0 onto Ω_t and is convecting the basis functions w_i we are concerned with.For any element y in $L^2(Q)$ let us define the transported element z in $L^2(Q_0)$ defined by $z = y\circ T_t$ which should be understood as $z(t,X) = y(t,T_t(X))$ .Before stating the result let us introduce some technical notations:

the derivative with respect to t of the transformation T_t verifies $\partial_t T_t = V(t) \circ T_t$ and it will be convenient to denote by M the following vector field which plays an essential role in the change of variable :

$$M(t) = (DT_t)^{-1} . \partial_t T_t = (DT_t)^{-1} . V(t) \circ T_t \qquad (3.1)$$

and we denote by M*M the symetric matrix defined by $M^*M .x = <M,x> M$

The Dalambertian operator associated to V will be defined by

$$\mathbb{H}(V).z = \partial^2_{tt} z - \mathrm{div}(\mathbb{A}(t) .\mathrm{grad}\, z) - [2\, \mathrm{grad}\, z_t . M(t) + \partial_t z\, R(t) + \mathrm{grad}\, z . S(t)] + J(t)\, |z|^{\beta} z \qquad (3.2)$$

where

$$\mathbb{A}(t) = (DT_t)^{-1} . ((DT_t)^*)^{-1} - M(t)^* M(t) \qquad (3.3)$$

$$J(t) = \det (DT_t) \qquad (3.4)$$

$$R(t) = J(t)^{-1} (\mathrm{div}(J(t)M(t) - \partial_t J(t)) \qquad (3.5)$$

$$S(t) = J(t)^{-1} [\partial_t (J(t)M(t)) + \mathbb{A}(t).\, \mathrm{grad}\, J(t)] \qquad (3.6)$$

Concerning the positiveness of the jacobian determinant we have the following result.

Proposition 7.

for any T,T>0 ,there exists a constant M depending on T and V,M>0 such that

$$J(t,x) \geq M \text{ for all } t, 0 \leq t \leq T \text{ and } x \text{ in } \Omega_o \qquad (3.7)$$

proof: we have J(0) = 1 and

$$J(t) = \exp[\int_o^t (\mathrm{div}\, V(s)) \circ T_s(V)\, ds \geq \exp[-TC] = M$$

where C is the norm in $L^\infty(Q_o)$ of the integrand which is continuous on Q_o .

Concerning the hyperbolicity of the operator $\mathbb{H}(\mathbb{V})$ we have from **G.Da Prato-J.P.Zolésio** [DPZ] we have the following result.

Proposition 8.

Assuming that the vector field V is hyperbolic: $|V(t,x)| \leq a < 1$ *for (t,x) in Q ,the matrix* $\mathbb{A}(t)(.)$ *is uniformely coercive on* cl (Q_o).

Proof: from (3.3) we get

$$< A(t,x).u,u> = | (DT_t^{-1})^*.u |^2 - (M(t,x).u)^2$$
$$= | (DT_t^{-1})^*.u |^2 -[<V(t) \circ T_t, (DT_t^{-1})^*.u>]^2$$
$$\geq | (DT_t^{-1})^*.u |^2 (1 - \| V(t) \circ T_t \|_{L^\infty(Q)})$$

But there exist k>0 such that ,for any (t,x) in $cl(Q_o)$

$$| (DT_t(x))^*.u | \leq k\, |u| \text{ then } | (DT_t^{-1})^*.u| \geq k^{-1} |u|$$

and finely we get

$$< A(t,x).u,u> \geq k^{-1} (1 - \| V(t) \circ T_t \|_{L^\infty(Q)}) \; |u|^2 \qquad \circ$$

Finely concerning the transported equation we have the following result.

Proposition 9.

The two following points are equivalwent:

i) *y belongs to* $W^{1,\infty}([0,T],L_2(\Omega_t)) \cap L^{\infty}([0,T],H_o^1(\Omega_t))$ *and verifies in the sens of* $D'(Q)'$

$$Hy = 0 \text{ in } Q,\ y|_{\Sigma} = 0,\ y(0,.)=y_o \text{ in } L^2(\Omega_0),\ \partial_t y(0,.)= y_1 \text{ in } [H^{1/2}(\Omega_o)_{00}]' \qquad (3.8)$$

ii) $z=y\circ T_t$ *belongs to* $W^{1,\infty}([0,T],L_2(\Omega_o)) \cap L^{\infty}([0,T],H_o^1(\Omega_o))$ *and verifies in the sens of* $D'(Q_o)'$

$$\mathbb{H}(\mathbb{V})\ z = 0 \text{ in } Q_o,\quad z|_{\Sigma o} = 0,\quad z(0,.)=y_o \text{ in } L^2(\Omega_0) \qquad (3.9)$$

$$\partial_t y(0,.)= y_1 + \text{grad } y_o .V(0) \text{ in } [H^{1/2}(\Omega_o)_{00}]' \qquad (3.10)$$

Proof: The first point i) derives from the following identities:

$$(\text{grad } y)\circ T_t = (DT_t{}^*)^{-1} . \text{grad } z$$

$$(\partial_t y)\circ T_t = \partial_t z - M(t). \text{grad } z \qquad (3.11)$$

For the second point we make of the following result.

Lemma l. Let μ and π be smooth test functions equal to zero on the boundaries of the evolution domain out of the basis $\{0\}x\Omega_o$ then the following identity holds :

$$\iint_Q Hy\ \mu\ dtdx + \int_{\Omega o} \partial_t y(0)\ \mu(0)\ dx$$
$$= \iint_{Qo} \mathbb{H}(\mathbb{V}) .z\ \pi\ dtdX + \int_{\Omega o} (\partial_t z(0)-V(0).\text{grad } z(0))\pi(0)\ dX$$

proof of the lemma:

then the change of variables in weak form leads to

$$\iint_Q (-\partial_t y\ \partial_t\mu + \text{grady grad } \mu)\ dtdx + ß \iint_Q |y|^p y\ \mu\ dtdx =$$

$$\iint_{Q_o} (-J(t)(\partial_t z - \langle \text{grad } z,M(t)\rangle)(\partial_t\pi - \langle \text{grad } \pi,M(t)\rangle)+\langle A(t).\text{gradz},\text{grad } \pi\rangle)\ dtdX$$
$$+\ ß \iint_{Qo} |z|^p z\ \pi\ J(t)\ dtdx$$

where $\pi=\mu\circ T_t$,

$$A(t) = J(t)\ (DT_t{}^*)^{-1}.DT_t{}^{-1}$$

Assuming that the test functions μ and π are smooth and equal to zero on the neiborhood,respectively,of $cl(\Sigma U(\{T\},\Omega_T))$ and $cl(\Sigma_o U(\{T\},\Omega_o))$,we obtain by part integration the following expressions.

$$\iint_Q (-\partial_t y\ \partial_t\mu + \text{grady grad } \mu)\ dtdx + ß \iint_Q |y|^p y\ \mu\ dtdx =$$
$$\iint_Q Hy\ \mu\ dtdx + \int_{\Omega o} \partial_t y(0)\ \mu(0)\ dx$$

and

$$\iint_{Q_o} \left(-J(t)(\partial_t z - \langle \text{grad } z, M(t)\rangle)(\partial_t \pi - \langle \text{grad } \pi, M(t)\rangle) + \langle \mathbb{A}(t).\text{grad} z, \text{grad } \pi\rangle \right) dtdX$$
$$+ \beta \iint_{Q_o} |z|^p z\, \pi\, J(t)\, dtdx$$
$$= \iint_{Q_o} \mathbb{H}(\mathbb{V}).z\, \pi\, dtdX + \int_{\Omega_o} (\partial_t z(0) - V(0).\text{grad } z(0))\pi(0)\, dX$$

3.2 Boundedness of the Approximation

Let be given an admissible field V and a dense family $W_1,\dots,W_m,\dots$ in $H_o^1(\Omega_o)$ and we consider an expension $z_m = z_m(t,x)$ of the following form :

$$z_m(t,x) = \sum_{1\le i\le m} h_{m,i}(t)\, W_i(x) \tag{3.12}$$

where the m functions $h_{m,i}$ are determined as being the solution of the following non linear system

$$\int_{\Omega_o} \mathbb{H}(\mathbb{V}).\, z_m\, W_j \quad dx = 0 \quad \text{for } 0<t<T, i=1,\dots,m \tag{3.13}$$

with the terms $h_{m,i}(0)$ and $h_{m,i}'(0)$ chosen such that the following convergences hold :

$$\sum_{i=1,..,m} h_{m,i}(0)W_i \text{ ----> } y_o \text{ in } H_o^1(\Omega_o)\,,\ \sum_{i=1,..,m} h_{m,i}'(0)W_i \text{ ----> } y_1 + \text{grad } y_o .V(0) \text{ in } L^2(\Omega_o)$$

as m -----> ∞

We then introduce the following energy term: with $z = z_m$ for short

$$2\,E_m(t) = \int_{\Omega_o} (\partial_t z\, \partial_t z + \langle \mathbb{A}(t).\text{grad } z, \text{grad } z\rangle)dx$$
$$+ \beta/p+1 \int_{\Omega_o} |z|^{p+1} dx \tag{3.14}$$

We get

$$E_m(t)' = \int_{\Omega_o} (\partial_t z\, \partial_{tt} z + \langle \mathbb{A}(t).\text{grad } z, \text{grad}\partial_t z\rangle + 1/2 \langle \partial_t \mathbb{A}(t) \text{ grad } z, \text{grad } z\rangle)dx$$
$$+ \beta \int_{\Omega_o} |z|^p\, \partial_t z\, dx \tag{3.15}$$

that is

$$E_m(t)' = \int_{\Omega_o} \mathbb{H}(\mathbb{V})\, z\, \partial_t z\, dx + 1/2 \int_{\Omega_o} \langle \partial_t \mathbb{A}(t) \text{ grad } z, \text{grad } z\rangle)dx +$$
$$+[2 \int_{\Omega_o} \text{grad } \partial_t z .M(t)\ \partial_t z\, dx + \int_{\Omega_o} R(t)\, \partial_t z\, \partial_t z\ dx + \int_{\Omega_o} \langle S(t).\text{grad } z\rangle \partial_t z\ dx\ ]$$

From (3.12) the first term in the right hand side vanishes; concerning the third one we use from the following identity

$$2 \int_{\Omega_o} < \text{grad}\, \partial_t z , M(t) > \partial_t z \, dx \; = \; \int_{\Omega_o} \text{grad}(\partial_t z . \partial_t z) M(t) \; dx$$

$$= - \int_{\Omega_o} \text{div}(M)(\partial_t z \, \partial_t z \;\;) dx \quad + \int_{\partial\Omega_o} \partial_t z \, \partial_t z \;\; M(t).n_o \, d\mu_o$$

where the last term vanishes for $\partial_t z$ is zero on the boundary .Finely one can find a constant C >0 such that

$E_m(t)' \le C \; E_m(t)$, for all t in [0,T] and all integer m .Then for all t in [0,T] and all integer m we do have

$$E_m(t) \le E_m(0) \;\; \exp(CT) \qquad (3.16)$$

From the assumptions concerning the terms $h_{m,i}(0)$ and $h_{m,i}(0)'$ we get the following convergence:

$$E_m(0) \text{ ----> } E(0) = 1/2 \int_{\Omega_o} (y_o y_o + y_1 y_1) \, dx \quad , \text{ as } m \text{---->} \infty \qquad (3.17)$$

From (3.15),(3.16) it follows that $E_m(t)$ remains bounded uniformely in t and m.From the uniform coercivity of the matrix $\mathbb{A}$ we obtain that the two sequences $\partial_t z_m$ and grad z_m remain bounded uniformely in t,$0 \le t \le T$,and $m \ge 0$ .Then one can find a subsequence z_m and an element z,both in $W^{1,\infty}([0,T],L_2(\Omega_o)) \cap L^\infty([0,T],H_o^{\;1}(\Omega_o))$ such that the following convergences hold.

$$\begin{aligned} & z_m \text{ -----> } z \text{ in } L^2(Q_o) \\ & \partial_t z_m \text{ -------> } \partial_t z \text{ weakly-}* \text{ in } L^\infty([0,T,L^2(\Omega_o)) \\ & \text{grad } z_m \text{ ------> grad } z \quad \text{weakly-}* \text{ in } L^\infty([0,T,L^2(\Omega_o)^n) \end{aligned} \qquad (3.18)$$

From Lemma 1 we get

$$0 = \int_{\Omega_o} \mathbb{H}(\mathbb{V}). \, z_m \;\; W_j \quad dx \; =$$

$$\iint_Q Hy_m w_j \, dtdx + \int_{\Omega o} \partial_t y_m(0) \; w_j(0) \, dx \; - \int_{\Omega o} (\partial_t z_m(0) - V(0).\text{grad } z_m(0)) W_j(0) \, dX \quad (3.19)$$

where we set

$$y_m(t,x) = \sum_{1 \le i \le m} h_{m,i}(t) \; w_i(t,x) \qquad \text{and } \; w_j(0) = w_j(0,.)$$

with $\qquad w_i(t,x) = W_i(T_t(V)^{-1}(x))$

in fact we have $y_m(t,.) = z_m(t,T_t(V)^{-1}(.))$

then as z_m satisfies the convergences (3.18) we obtain the analougous convergences for the sequence y_m to an element $y = z \circ T_t(v)^{-1}$.

3.3 The limiting element y is solution to the problem (1.16)-(1.20)

We consider

$S=\{q(t,x), q=\sum h_i(t)\, w_i(t,x)\ /\ h_i$ belongs to $C^\infty([0,T])$ with $h_i(0)=h_i(T)=0\}$

it can be easily verified that S is dense in $H_0^1(Q)$ (for $S_o=\{q_o=q\circ T_t\ /\ q$ in S $\}$ is classiquely known to be dense in $H_o^1(Q_o)$) .From (3.12) and the Green's first theorem we get :

$\iint_Q \text{grad } y_m.\text{grad } q - \partial_t y_m\, \partial_t q\ dtdx = \iint_Q Hy_m\, q\, dt\, dx = 0$

but the left hand side converges to

$\iint_Q \text{grad } y.\text{grad } q - \partial_t y\, \partial_t q\ dt\, dx$ which is equal by Green's theorem to $\iint_Q H.y\, q\, dt\, dx$,then we get by density of S, $\iint_Q H.y\, q\, dt\, dx = 0$ for all q in $H_0^1(Q)$,that is Hy=0.The initial conditions on y

follow from the particular choices of $h_{m,i}(0)$ and $\partial_t h_{m,i}(0)$ we did.To end with the the proof of theorem 1 one have just now to verify that the conditions (3.11)-(3.12) which define the sequence z_m are equivalent after change of variable T_t to (1.22) -(1.8) which determines the sequence y_m.

Remark 4. For several purpose we face to the characterization of second order shape derivatives, see M.C.Delfour-J.P.Zolésio[DZ].Here the acceleration $\partial_t V$ plays a central role.

References.

[L]J.L.Lions. Une remarque sur les problemes d'evolution nonlineaires dans les domaines non cylindriques,Rev.Roum.Pure Appl.9(1964),11-18.

[Z1] J.P.Zolésio. Approximation for Wave Equation in Non Cylindrical Domain.Proc. Comcon Conf."stabilization of Flexible Structures", Montpellier 1989,Springer Verlag LNCIS,to appear

[Z2] J.P.Zolésio. Stabilization of Wave Equation by Periodical Moving Actuators In proc. IFAC Fith Symposium "Control of Distributed Parameter Systems" El Jai,Amouroux eds.,Perpignan 1989.

[CS] J.Cooper,W.A. Strauss. Energy Boundedness and Decay of Wave Reflecting off a Moving Obstacle,Indiana Univ.Math.J.,25(1976),pp. 671-690

[CB] J.Cooper,C. Bardos. A non linear wave equation in non cylindrical domain.J.of Mathatical Analysis and Applications 38-60 (1973)

[BC] C.Bardos,G. Chen. Control and Stabilization for the Wave Equation.Part III. SIAM J. on Control and Optim. vol 19,n°1,january 1981.

[DPZ1] G.Da Prato,J.P.Zolésio.Existence and Control for Wave Equation in Non Cylindrical Domains.in proc. of Comcon conf. Montpellier 1989,Springer Verlag 1990,LNCIS,to appear.

[DPZ2] G.Da Prato,J.P.Zolésio. A Boundary Control Problem for a Parabolic Equation in Non cylindrical Domain.A.V. Balakrishnan,J.P.Zolésio eds., "Stabilization of Flexible Structures",Optimization Software inc.,L.A., N.Y.,1988

[DPZ3] G.Da Prato,J.P.Zolésio.Dynamical Programing for Non Cylindrical Parabolic Equation.S.& C. Letters,11(1988) n°1pp.73-77.

[CDZ1] P.Cannarsa,G.Da Prato,J.P.Zolésio.The Damped WAve Equation in Moving Domain, J.Diff. Equation,to appear

[CDZ2] P.Cannarsa,G.Da Prato,J.P.Zolésio.Dynamical Actuators for the Heat Equation. Proc.IFAC Symposium,perpignan 1989

[CDZ3] P.Cannarsa,G.Da Prato,J.P.Zolésio.Dynamical Shape Control of the Heat Equation ,S.&C. Letters

[CDZ4]P.Cannarsa,G.Da Prato,J.P.Zolésio.Evolution Equations in non cylindrical domains. Rendiconti Lincei

[CDZ5]P.Cannarsa,G.Da Prato,J.P.Zolésio.Riccati Equation in Non Cylindrical Domains. In Proc. of Comcon Conf. Montpellier 1989,Spriger Verlag ,LNCIS 1990

[SI] J.Sikorav C.R.Acad.Sc. Paris, 1989

[Z3] J.P.Zolésio. Un Résultat d'Existence de Vitesse Convergente en Optimisation de Domaine.C.R.Acad.Sc.,Paris 283 (1976),Ser. A,Pp. 855-859

[Z4] J.P.Zolésio. The Material Derivative (or Speed Method) for Shape Optimization.in "Optimization of Distributed Parameter Systems", J.Céa,E.Haug eds.,SIjthoff aand Noordhoff(1980),pp.1089-1194..

[Z5] J.P.Zolésio.Boundary Control and Boundary Variations ,LNCIS n 100,Springer Verlag,1987

[SZ]J.Sokolowski,J.P.Zolésio.Shape Sensitivity Analysis of Unilateral Problems,SIAM J.on Math. Anal.18(1987).

[DZ] M.C.Delfour,J.P.Zolésio . to appear in SIAM J. on Control and Optim.

NUMERICAL ANALYSIS

M CHIPOT

Numerical analysis of oscillations in two wells problems

Université de Metz
Département de Mathématiques
Ile de Saulcy, 57045 METZ-CEDEX 01
(FRANCE)

1. Introduction

The goal of this note is to explain, in a simple case, new techniques developed recently regarding the numerical analysis of nonconvex problems (see [C.L.$_1$]-[C.L.$_4$], [C.C.], [C.$_2$]). In general such problems fail to have a minimizer. However, minimizing sequences tend to organize themselves according, for instance, to the position with respect to the wells of the boundary data that has to be matched. Such a physical situation is encountered for instance in the theory of hyperelasticity for structured materials as crystals (see [B.J.$_1$], [B.J.$_2$], [C.$_1$] [Er.], [F.$_1$], [F.$_2$], [J.K.], [K.], [Ko.], [S.]).

Let Ω be a polygonal domain of $\mathbf{R}^n$ of boundary Γ, $n \geq 1$. Let $\varphi : \mathbf{R}^n \to \mathbf{R}$ be a continuous function for which there exist $W_1 \neq W_2$ such that

$$\varphi(W_1) = \varphi(W_2) = 0 \quad , \quad \varphi(W) > 0 \quad \forall \quad W \notin \{W_1, W_2\}. \tag{1.1}$$

Set

$$A = \lambda W_1 + (1-\lambda) W_2 \quad , \quad \lambda \in (0,1). \tag{1.2}$$

For $v \in H^1(\Omega)$ we define two energy functionals by

$$E_1(v) = \int_\Omega \varphi(\nabla v(x))\, dx \tag{1.3}$$

$$E_2(v) = \int_\Omega \varphi(\nabla v(x)) + (v(x) - A \cdot x)^2\, dx \tag{1.4}$$

($A \cdot x$ denotes the scalar product of A and x and we make the convention that E_1 and E_2 are equal to $+\infty$ if $\varphi(v(x)) \notin L^1(\Omega)$ - this avoids to impose some growth condition on φ).

Let us denote by $H^1_A(\Omega)$ the set

$$H^1_A(\Omega) = \{v \in H^1(\Omega) \mid v(x) = A \cdot x \quad \text{on} \quad \partial\Omega\}.$$

Then, we would like to consider the following model problems

$$\inf_{H^1_A(\Omega)} E_1(v) \tag{P_1}$$

$$\inf_{H^1_A(\Omega)} E_2(v) \tag{P_2}$$

$$\inf_{H^1(\Omega)} E_2(v) \tag{P_3}.$$

(We do not consider $\inf_{H^1(\Omega)} E_1(v)$ which admits as trivial minimizers the functions $x \to W_i \cdot x$).

When we write "inf" in (P_i) we have in mind different questions: first compute the infimum, then see if there exists a minimizer, finally, if not, study the minimizing sequences. In the present case we will see that there is no minimizer for (P_i) and so we will investigate more closely the behaviour of the minimizing sequences.

2. Energy estimate

First, note that without loss of generality we can assume that $A = 0$. Indeed if $v \in H^1_A(\Omega)$, then $\tilde{v} = v - A \cdot x \in H^1_0(\Omega)$ and

$$\begin{aligned} \int_\Omega \varphi(\nabla v(x))\, dx &= \int_\Omega \varphi(\nabla \tilde{v}(x) + A)\, dx \\ &= \int_\Omega \tilde{\varphi}(\nabla \tilde{v}(x))\, dx \end{aligned} \tag{2.1}$$

where we have set

$$\tilde{\varphi}(W) = \varphi(W + A).$$

Thus, for instance to minimize E_1 on $H^1_A(\Omega)$ is equivalent to minimize the last integral of (2.1) on $H^1_0(\Omega)$. Thus, from now on in this section, we will assume that $A = 0$.

Let us show that the infima of (P_i) are all equal to 0. For that let us consider a triangulation τ of Ω with simplices of diameters less then h. If K is some simplex of τ denote by $P_1(K)$ the space of polynomials of degree 1 on K and set

$$V_h = \{v : \Omega \to \mathbf{R}, \;\; \text{continuous} \mid v_{|K} \in P_1(K) \;\; \forall\, K \in \tau\}$$

$$V_h^0 = \{v \in V_h \mid v = 0 \;\; \text{on} \;\; \Gamma\}$$

($v_{|K}$ denotes the restriction of v to K).

Then we have

THEOREM 1: There exists a constant C such that

$$\inf_{V_h^0} E_1(v) \text{ or } E_2(v) \leq C \cdot h^{1/2} \tag{2.2}$$

$$\inf_{V_h} E_2(v) \leq C \cdot h^{2/3}. \tag{2.3}$$

Proof: We are interested in proving (2.2), (2.3) when h is small i.e. $h < 1$.

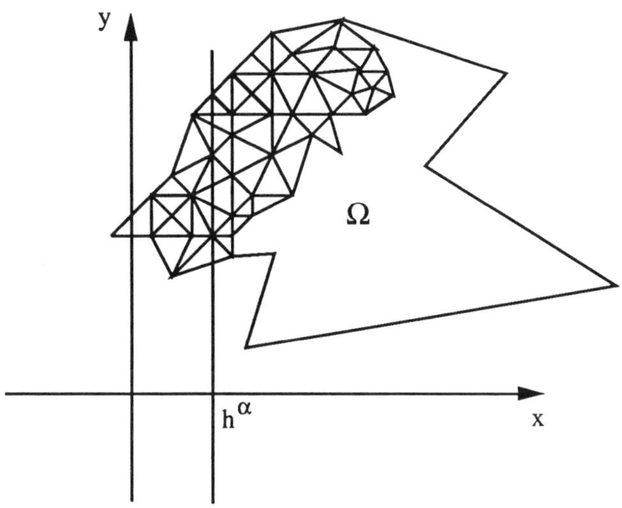

For simplicity let us assume that $n = 2$. Furthermore, let us assume that

$$W_1 = e_1 \quad , \quad W_2 = -e_1 \tag{2.4}$$

where e_1 is the first vector of the canonical basis of $\mathbf{R}^2$. It is very easy to reduce the general case to this one. We refer the reader to [C.$_2$], [C.C.], [C.C.K.], for generalizations when $n > 2$ and when the number of wells is arbitrary. However, note that the basic idea of the proof is contained in the two dimensional case presented here.

For any $\alpha \in (0,1)$ one has

$$h^\alpha >> h$$

and the vertical lines $x = k.h^\alpha$, $k \in \mathbf{Z}$ are dividing Ω in different strips (see the above figure).

We define u_h as the function of x only, periodic of period $2h^\alpha$ and such that

$$u_h(x,y) = \begin{cases} x & \text{for} \quad 0 \leq x \leq h^\alpha \\ -x + 2h^\alpha & \text{for} \quad h^\alpha \leq x \leq 2h^\alpha. \end{cases}$$

One has clearly

$$0 \leq u_h \leq h^\alpha. \tag{2.5}$$

Moreover, u_h is affine on each triangle except one those crossed by a line $x = k.h^\alpha$. On these triangles we modify u_h interpolating linearly - i.e. we replace u_h by the function of $P_1(K)$ which agrees with u_h on the vertices of K. We still denote by u_h the modified function. Of course (2.5) is preserved and the gradient of u_h is uniformly bounded. So we have (recall that $A = 0$):

$$\begin{aligned} E_2(u_h) &= \int_\Omega \varphi(\nabla u_h(x)) + u_h(x)^2 \, dx \\ &\leq \int_S \varphi(\nabla u_h(x)) \, dx + |\Omega|.h^{2\alpha} \end{aligned}$$

where S is the set of triangles where $\nabla u_h \neq \pm e_1$, $|\Omega|$ the area of Ω.

Since $\varphi(\nabla u_h)$ is bounded by some constant C we get

$$E_2(u_h) \leq C|S| + |\Omega|.h^{2\alpha}.$$

The diameter of each triangle is less than h so we have

$$|S| \leq 2N.dh$$

where N is the number of lines $x = k.h^{2\alpha}$ crossing Ω and d is the diameter of Ω. Since

$$(N-1)h^\alpha \leq d \quad \Rightarrow \quad N.h^\alpha \leq d+1$$

we obtain for some constant C

$$\begin{aligned} E_2(u_h) &\leq C.h^{1-\alpha} + |\Omega|.h^{2\alpha} \\ &\leq C.h^{min(1-\alpha, 2\alpha)}. \end{aligned}$$

But $min(1-\alpha, 2\alpha)$ is maximum when $1 - \alpha = 2\alpha$ - i.e. when $\alpha = 1/3$ and (2.3) follows. When we want to minimize on V_h^0 considering u_h defined above we modify it once more replacing it by

$$min(u_h, dist(., \Gamma))$$

where $dist(., \Gamma)$ denotes the distance of x to Γ, or more precisely by the interpolate of this last function - i.e. by the function of V_h^0 which agrees with $min(u_h, dist(., \Gamma))$ on the nodes of the triangulation. Then clearly, if we still denote by u_h this new function, one has

$$\nabla u_h(x) = \pm e_1 \quad \text{a.e. in} \quad \Omega$$

except on a set S union of a set located around the lines $x = k.h^\alpha$ crossing Ω and an other one within a distance less then h^α of Γ (see (2.5)). Thus if $|S|$ denotes the area of S and $|\Gamma|$ the length of Γ

$$|S| \le C.h^{1-\alpha} + |\Gamma|.h^\alpha.$$

Hence arguing as above we obtain for this u_h

$$E_1(u_h) \le E_2(u_h) \le C.(h^{1-\alpha} + h^\alpha + h^{2\alpha})$$

$$\le C.h^{min(1-\alpha,\alpha)}.$$

But $min(1-\alpha, \alpha)$ is maximum when $1-\alpha = \alpha$ i.e. when $\alpha = 1/2$ which gives (2.2) and concludes the proof.

Remark 1: It is clear that (2.2), (2.3) imply that the infimum of each of the problems (P_i) is zero. Recall that we have assumed that $A = 0$ and note that $V_h \subset H^1(\Omega)$ (Resp. $V_h^0 \subset H_0^1(\Omega)$). This is also of course the consequence of the relaxation theory.

Remark 2: We do not know if these estimates are sharp. We refer to [C.L.$_4$], [C.$_2$] for results in this direction.

3. Nonexistence of minimizer - Young measure

If for each of the problems (P_i) the infimum is 0 none of them admits a minimizer.

Indeed we have

THEOREM 2: The problems (P_i) do not admit a minimizer.

Proof: First consider the case of $(P_2), (P_3)$. If u is a minimizer then

$$\int_\Omega \varphi(\nabla u(x)) + (u(x) - A \cdot x)^2 \, dx = 0$$

hence by (1.1)

$$u(x) = A \cdot x \quad , \quad \nabla u(x) = W_1 \text{ or } W_2 \quad \text{a.e. in} \quad \Omega.$$

But since $A \notin \{W_1, W_2\}$ this is impossible.

In the case of (P_1) if u is a minimizer one has

$$\int_\Omega \varphi(\nabla u(x)) \, dx = 0$$

hence by (1.1)

$$\nabla u(x) = W_1 \quad \text{or } W_2 \quad \text{a.e. in} \quad \Omega. \tag{3.1}$$

Let ν be a unit vector orthogonal to $W_1 - A$ (hence also to $W_2 - A$ by (1.2)). From (3.1) we deduce

$$\frac{\partial}{\partial \nu}(u(x) - A \cdot x) = (\nabla u(x) - A) \cdot \nu = 0 \quad \text{a.e. in} \quad \Omega.$$

Hence, since $u(x) - A \cdot x = 0$ on Γ we get

$$u(x) = A \cdot x \quad \text{in} \quad \Omega$$

which again contradicts (3.1).

In the absence of minimizer we turn to the study of minimizing sequences. As we will see they have a very precise pattern. This is explained by the fact that there exists a unique Young measure associated to the problems at hand. Indeed, first let us denote by $|\ |_\infty$ the usual norm in $L^\infty(\Omega)$, $|\ |$ the Euclidean norm. Recall that if u_h is a minimizing "sequence" of (P_i) and if for some constant C

$$|u_h|_\infty \quad , \quad ||\nabla u_h||_\infty \leq C \tag{3.2}$$

then u_h defines a parametrized measure ν_x on $\mathbf{R}^n$ -so called Young measure- in the sense that there exists a subsequence of u_h, that we still will denote by u_h, and a probability measure ν_x such that for any bounded Carathéodory function $F(x, \xi)$ one has

$$\lim_{h \to 0} \int_\Omega F(x, \nabla u_h(x)) \, dx = \int_\Omega \int_{\mathbf{R}^n} F(x, \lambda) \, d\nu_x(\lambda) dx \tag{3.3}$$

(see [B.], [C.K.], [K.], [K.P.$_1$], [K.P.$_2$], [D.], [Ev.], [T.], [V.], for these questions).

Then we can prove

THEOREM 3: There exists a unique Young measure associated to the problems (P_i) which is given by

$$\nu_x = \lambda \delta_{W_1} + (1 - \lambda) \delta_{W_2} \tag{3.4}$$

where δ_{W_i} denotes the Dirac measure at W_i, λ the parameter of (1.2).

Proof: Let u_h be a minimizing sequence of (P_i) such that (3.2) holds. According for instance to [T.] u_h defines a Young measure ν_x. Applying (3.3) with $F(x, \xi) = \varphi(\xi)$ we get by Theorem 1

$$\int_\Omega \varphi(\nabla u_h(x)) dx \to \int_\Omega \int_{\mathbf{R}^n} \varphi(\lambda) d\nu_x(\lambda) dx = 0.$$

From (1.1) it results that ν_x is supported by $W_1, \ W_2$. Thus, since ν_x is a probability measure one has

$$\nu_x = \beta(x) \delta_{W_1} + (1 - \beta(x)) \delta_{W_2} \tag{3.5}$$

where $\beta(x)$ is a function taking its values in $(0,1)$. One can extract a subsequence of u_h still denoted by u_h such that for some $u \in W^{1,\infty}(\Omega)$ (see [G.T.] for references on Sobolev spaces)

$$u_h \to u \quad \text{uniformly in } \Omega \tag{3.6}$$

$$\nabla u_h \rightharpoonup \nabla u \quad \text{in } L^\infty(\Omega) - \text{weak} * . \tag{3.7}$$

Let us consider first the case of $(P_2), (P_3)$.

Since $E_2(u_h) \to 0$ one has $u_h \to A \cdot x$ in $L^2(\Omega)$ hence necessarily

$$u(x) = A \cdot x. \tag{3.8}$$

Since (3.7), (3.8) hold for any converging subsequence one has

$$\nabla u_h \rightharpoonup A \quad \text{in } L^\infty(\Omega) - \text{weak} * .$$

Applying then (3.3) with $F(x,\xi) = \chi_B(x)\xi$ where B is any ball included in Ω it follows that

$$\int_B \nabla u_h(x) dx \to \int_B A dx = \int_\Omega \int_{\mathbf{R}^n} \chi_B(x) \lambda d\nu_x(\lambda) dx.$$

Hence by (3.5)

$$A = \frac{1}{|B|} \int_B \beta(x) dx . W_1 + \frac{1}{|B|} \int_B (1 - \beta(x)) dx . W_2$$

and by the uniqueness of barycentric coordinates

$$\lambda = \frac{1}{|B|} \int_B \beta(x) dx.$$

Since this holds for any ball B included in Ω, by the Lebesgue differentiation theorem we obtain

$$\beta(x) = \lambda \qquad \text{a.e. } \; x \in \Omega$$

and (3.4) follows.

In the case of (P_1) we denote by ν a unit vector orthogonal to $W_i - A$. If we apply (3.3) with

$$F(x,\xi) = |(\xi - A) \cdot \nu|$$

we obtain

$$\int_\Omega |(\nabla u_h(x) - A) \cdot \nu| dx \to \int_\Omega \int_{\mathbf{R}^n} |(\lambda - A) \cdot \nu| d\nu_x(\lambda) dx = 0$$

by (3.5). It results that

$$\frac{\partial}{\partial \nu}(u_h - A \cdot x) \to 0 \text{ in } L^1(\Omega).$$

Then by the Poincaré inequality

$$u_h(x) \to A \cdot x \text{ in } L^1(\Omega).$$

Hence (3.8) holds and one concludes as in the case of $(P_2), (P_3)$.

4. Numerical Analysis of oscillations

The uniqueness of the Young measure associated to (P_i) is an indication of some coherence in the oscillations of the minimizing sequences. This is what we would like to investigate now.

For this let

$$0 < R < \frac{1}{2}|W_1 - W_2|.$$

Let B be a ball included in Ω. For $v \in H^1(\Omega)$, set

$$B_i^R(v) = B_i^R = \{x \in B \mid |\nabla v(x) - W_i| < R\}.$$

If $|\ |$ denotes the Lebesgue measure, then $|B_i^R(v)|/|B|$ is the probability for v to have its gradient in the ball of center W_i and radius R. Let us denote by $B_{ex}^R(v)$

$$B_{ex}^R(v) = B_{ex}^R = B \backslash \{B_1^R \cup B_2^R\}.$$

Let π be the mapping defined by

$$\pi(\xi) = \begin{cases} W_1 & \text{if } |\xi - W_1| \le |\xi - W_2| \\ \\ W_2 & \text{if } |\xi - W_1| > |\xi - W_2| \end{cases} \tag{4.1}$$

and assume that there exists a constant $\wedge$ such that

$$\begin{aligned} \varphi(\xi) &\ge \wedge.|\xi - \pi(\xi)|^2 \\ \\ &= \wedge.min(|\xi - W_1|^2, |\xi - W_2|^2) \end{aligned} \tag{4.2}$$

Then we have

THEOREM 4: Assume that φ satisfies (4.2). There exists a constant C such that

$$|\frac{|B_1^R(v)|}{|B|} - \lambda| \le C.E_1(v)^{\frac{1}{2}} \tag{4.3}$$

$$\left|\frac{|B_2^R(v)|}{|B|} - (1-\lambda)\right| \le C.E_1(v)^{\frac{1}{2}} \tag{4.4}$$

Proof: First remark that

$$|B_1^R|.(W_1 - A) + |B_2^R|.(W_2 - A) = \int_{B_1^R \cup B_2^R} (\pi(\nabla v(x)) - A)\, dx$$

$$= \int_B (\pi(\nabla v(x)) - A)\, dx - \int_{B_{ex}^R} (\pi(\nabla v(x)) - A)\, dx$$

$$= \int_B (\pi(\nabla v(x)) - \nabla v(x))\, dx + \int_B (\nabla v(x) - A)\, dx - \int_{B_{ex}^R} (\pi(\nabla v(x)) - A)\, dx$$

$$\stackrel{\text{def}}{=} I_1 + I_2 + I_3. \tag{4.5}$$

To estimate I_1 remark that by Hölder's inequality and (4.2)

$$\begin{aligned} |I_1| &\le \int_B |\pi(\nabla v(x)) - \nabla v(x)|\, dx \\ &\le |B|^{\frac{1}{2}} \left(\int_B |\pi(\nabla v(x)) - \nabla v(x)|^2\, dx \right)^{\frac{1}{2}} \\ &\le \wedge^{-\frac{1}{2}} |B|^{\frac{1}{2}} E_1(v)^{\frac{1}{2}}. \end{aligned} \tag{4.6}$$

To estimate I_2 one applies the divergence theorem to get

$$I_2 = \int_B (\nabla v(x) - A)\, dx = \int_{\partial B} (v(x) - A \cdot x)\, n(x)\, d\sigma(x)$$

where $n(x)$ denotes the outward normal to ∂B, $d\sigma(x)$ the usual surface measure on ∂B. Hence :

$$\begin{aligned} |I_2| &\le \int_{\partial B} |v(x) - A \cdot x|\, d\sigma(x) \\ &\le |\partial B|^{\frac{1}{2}} \left(\int_{\partial B} |v(x) - A \cdot x|^2\, d\sigma(x) \right)^{\frac{1}{2}} \\ &\le |\partial B|^{\frac{1}{2}} \left(\int_B |v(x) - A \cdot x|^2 + |\nabla v(x) - A|^2\, dx \right)^{\frac{1}{2}} \\ &\le C \left(\int_B |\frac{\partial}{\partial \nu}(v(x) - A \cdot x)|^2\, dx \right)^{\frac{1}{2}} \end{aligned} \tag{4.7}$$

by Poincaré's inequality and if ν denotes any unit vector. Let us choose ν orthogonal to $W_i - A$. Then (4.7) reads

$$|I_2| \le C \left(\int_B |P_\nu(\nabla v(x) - A)|^2\, dx \right)^{\frac{1}{2}} \tag{4.8}$$

where P_ν denotes the orthogonal projection on the subspace of $\mathbf{R}^n$ spanned by ν. From (4.2) we deduce

$$\varphi(\xi) \geq \wedge . min(|\xi - W_1|^2, |\xi - W_2|^2)$$

$$\geq \wedge . min(|(\xi - A) - (W_1 - A)|^2, |(\xi - A) - (W_2 - A)|^2)$$

$$\geq \wedge . |P_\nu(\xi - A)|^2.$$

Hence (4.8) becomes

$$|I_2| \leq C.E_1(v)^{\frac{1}{2}} \tag{4.9}$$

To estimate I_3, note that

$$|I_3| \leq \int_{B^R_{ex}} |\pi(\nabla v(x)) - A|\, dx$$
$$\leq \operatorname*{Max}_i |W_i - A|\ |B^R_{ex}|.$$

Now, clearly

$$R|B^R_{ex}| \leq \int_{B^R_{ex}} |\pi(\nabla v(x)) - \nabla v(x)|\, dx$$
$$\leq \int_B |\pi(\nabla v(x)) - \nabla v(x)|\, dx.$$

Hence, using Hölder's inequality

$$|B^R_{ex}| \leq \frac{1}{R} |B|^{\frac{1}{2}} \wedge^{-\frac{1}{2}} E(v)^{\frac{1}{2}} \tag{4.10}$$

and thus

$$|I_3| \leq C.E_1(v)^{\frac{1}{2}}. \tag{4.11}$$

Combining (4.5), (4.6), (4.9), (4.11) we obtain

$$|\frac{|B^R_1|}{|B|}.(W_1 - A) + \frac{|B^R_2|}{|B|}.(W_2 - A)| \leq C.E_1(v)^{\frac{1}{2}}.$$

Remark that $\frac{|B^R_2|}{|B|} = 1 - \frac{|B^R_1|}{|B|} - \frac{|B^R_{ex}|}{|B|}$. Since

$$\lambda(W_1 - A) + (1 - \lambda)(W_2 - A) = 0$$

we deduce by subtraction and by (4.10)

$$|\lambda - \frac{|B^R_1|}{|B|}||W_1 - W_2| \leq C.E_1(v)^{\frac{1}{2}}$$

and the result follows since a similar estimate holds for $|(1-\lambda)-\frac{|B_2^R|}{|B|}|$.

As a consequence we see that if u_h is some function of V_h or V_h^0 such that

$$E_1(u_h) \leq 2.\inf_{V_h^0} E_1(v)$$

or

$$E_2(u_h) \leq 2.\inf_{V_h^0} E_2(v)$$

or

$$E_2(u_h) \leq 2.\inf_{V_h} E_2(v)$$

then by Theorem 1 one has

$$|\frac{|B_1^R(u_h)|}{|B|} - \lambda| \leq C.h^{\frac{1}{4}}$$

$$|\frac{|B_2^R(u_h)|}{|B|} - (1-\lambda)| \leq C.h^{\frac{1}{4}}$$

in the two first cases and

$$|\frac{|B_1^R(u_h)|}{|B|} - \lambda| \leq C.h^{\frac{1}{3}}$$

$$|\frac{|B_2^R(u_h)|}{|B|} - (1-\lambda)| \leq C.h^{\frac{1}{3}}.$$

in the third one. This gives an estimation in terms of h of the probability for u_h to have its gradient in a neighbourhood of W_i. This probability approaches λ or $1-\lambda$ when $h \to 0$.

We would like to end this section by some numerical results. Let us assume that Ω is the square $(-1,1)^2$, and consider the case of the two wells

$$W_1 = (-1,1) \quad , \quad W_2 = (1,-1).$$

Moreover let us choose

$$\varphi(\xi) = |\xi - W_1|^2 \; |\xi - W_2|^2.$$

It is not difficult to check that these data fit our previous analysis. Since

$$0 = \frac{1}{2}W_1 + \frac{1}{2}W_2$$

the minimizing sequences of (P_i) are choosing their gradients alternatively around W_1 and W_2 with a probability which tends to $\frac{1}{2}$ (see Theorem 4). This can be observed on the figure below in the case of (P_1):

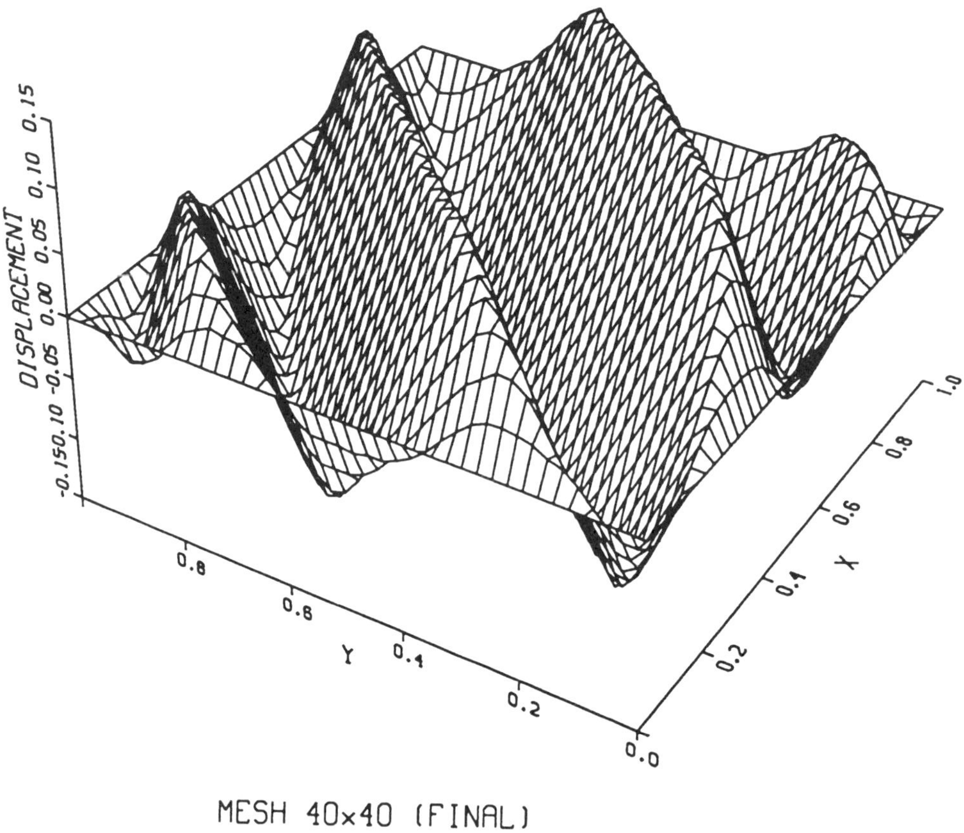

Computational Result for $h = 1/40$.

In fact, in order to lower the energy, the computer uses almost exactly the technique developed in Theorem 1. Of course, the oscillations of the minimizing sequences are stopped at the level of the grid.

Our numerical codes use standard iterations algorithms. We start with a random distribution of initial values of the function close to 0 and we decrease the energy using a conjugate gradient method. An example of the final result is displayed on the above figure.

The most interesting examples of applications of our results occur in the study of deformations of ordered materials as crystals. In this case the potential wells are matrices. Moreover, due to the frame indifference, these wells are in infinite number. Also deformations have to satisfy the compatiblity condition. This makes the problem more involved. For some results in this direction and complements we refer the reader to [C.C.K.] and [Co.].

Acknowledgements: This work was done during one visit at the University of Minnesota. I would like to thank the I.M.A. and the School of Mathematics who are providing such a good working atmosphere. This research has been supported there by the NSF and ASOFR through DMS 87-18881.

I thank also C. Collins and D. Kinderlehrer for several fruitful discussions on the subject.

REFERENCES

[B.] J. M. Ball : A version of the fundamental theorem for Young measures. In *Partial Differential Equations and Continuum Models of Phase Transitions*, Lecture Notes in Physics # 344, M. Rascle, D.Serre and M. Slemrod, eds., Springer-Verlag, (1989), p. 207-215.

[B.J.$_1$] J. M. Ball and R. D. James : Fine phase mixtures as minimizers of energy. Arch. Rational Mech. Anal., 100, (1987), p. 13-52.

[B.J.$_2$] J. M. Ball and R. D. James : Experimental tests of a theory of fine microstructures. (1988), preprint.

[B.M.] J. M. Ball and F. Murat : $W^{1,p}$ quasiconvexity and variational problems for multiple integrals. J. Funct. Anal., 58, (1984), p. 255-253.

[C.$_1$] M. Chipot : Hyperelasticity for crystals. European Journ. of Appl. Math., 1, (1990), p. 113-129.

[C.$_2$] M. Chipot : Numerical analysis of oscillations in nonconvex problems. (To appear).

[C.C.] M. Chipot and C. Collins : Numerical approximation in variational pro blems with potential wells. (1990), preprint. (To appear).

[C.C.K.] M. Chipot, C. Collins and D. Kinderlehrer : (In preparation).

[C.K.] M. Chipot and D. Kinderlehrer : Equilibrium configurations of crystals. Arch. Rational Mech. Anal., 103, (1988), p. 237-277.

[Co.] C. Collins : Thesis, University of Minnesota, 1990.

[C.K.L.] C. Collins, D. Kinderlehrer and M. Luskin : Numerical approximation of the solution of a variational problem with a double well potential. IMA preprint # 605, (1990). (To appear in SIAM J. Numer. Anal.).

[C.L.$_1$] C. Collins and M. Luskin : The computation of the austenitic-martensitic phase transition. In *Partial Differential Equations and Continuum Models of Phase Transitions*, Lecture Notes in Physics # 344, M. Rascle, D.Serre and M. Slemrod, eds., Springer-Verlag, Berlin, New York, (1989),

p. 34–50.

[C.L.$_2$] C. Collins and M. Luskin : Computational results for phase transitions in shape memory materials.
In *Smart Materials, Structures, and Mathematical Issues*, C. Rogers, ed. Technomic Publishing Co., Lancaster, Pennsylvania, (1989), p. 198–215.

[C.L.$_3$] C. Collins and M. Luskin : Numerical modeling of the microstructure of crystals with symmetry-related variants.
In *Proceedings of the ARO US-Japan Workshop on Smart/ Intelligent Materials*, Technomic Publishing Co., Lancaster, Pennsylvania, (to appear).

[C.L.$_4$] C. Collins and M. Luskin : Optimal order error estimates for the finite elementapproximation of the solution of a nonconvex variational problem. Preprint, (1990).

[D.] B. Dacorogna : *Weak continuity and weak lower semicontinuity of nonlinear functionals.*
Springer Lecture Notes # 922, (1982).

[Er.] J. L. Ericksen : Some constrained elastic crystals.
In *Material Instabilities in Continuum Mechanics and Related Problems*, J. M. Ball, ed., Oxford University Press, Oxford, (1987), p. 119-137.

[Ev.] L. C. Evans : *Weak Convergence Methods for Nonlinear Partial Differential Equations.*
A.M.S. Regional Conference Series in Mathematics # 74, (1989).

[F.$_1$] I. Fonseca : Variational methods for elastic crystals.
Arch. Rat. Mech. Anal., 97, (1985), p. 189-220.

[F.$_2$] I. Fonseca : The lower quasiconvex envelope of the stored energy function for an elastic crystal.
J. Math. Pures et Appl., 67, (1988), p. 175–195.

[G.T.] D. Gilbarg and N.S. Trudinger : *Elliptic Partial Differential Equations of Second Order.*
Springer Verlag, (1985).

[J.$_1$] R. James : Basic principles for the improvement of shape-memory and related materials.
In *Smart Materials, Structures, and Mathematical Issues*, C. Rogers, ed., Technomic Publishing Co. Lancaster, PA, (1989).

[J.$_2$] R. D. James : Microstructure and weak convergence.
In *Material Instabilities in Continuum Mechanics and Related Pro blems*, J. M. Ball, ed., Oxford University Press, Oxford , (1987), p. 175-196.

[J. K.] R. D. James and D. Kinderlehrer : Theory of diffusionless phase transi tions.

In *Partial Differential Equations and Continuum Models of Phase Transitions*, Lecture Notes in Physics # 344, M. Rascle, D.Serre and M. Slemrod, eds., Springer-Verlag, Berlin, New York, (1989), p. 51-84

[K.] D. Kinderlehrer : Remarks about equilibrium configurations of crystals. In *Material Instabilities in Continuum Mechanics and Related Pro blems*, J. M. Ball ed., Oxford University Press, Oxford, (1987), p. 217-242.

[K. P.$_1$] D. Kinderlehrer and P. Pedregal : Analysis of Young measures (Preprint).

[K. P.$_2$] D. Kinderlehrer and P. Pedregal : Weak convergence of integrands and the Young measure representation (Preprint).

[Ko.] R. Kohn : The relationship between linear and nonlinear variational models of coherent phase transitions. In *Proceedings of the Seventh Army Conference on Applied Mathematics and Computing*, West Point, June 1989.

[P.] P. Pedregal : Thesis, University of Minnesota, 1989.

[S.] S. A. Silling : Phase changes induced by deformation in isothermal elastic crystals. (1988), (preprint).

[T.] L. Tartar : Compensated compactness and application to partial differential equations In *Nonlinear analysis and mechanics: Heriot–Watt Symp IV*, R.J. Knops ed., Pitman, (1979), 136–212.

[V.] M. Valadier : Young measures. CIME Course (1989), preprint.

B KLARES

Numerical invariants and Jordan form of a matrix

Abstract
We give numerical invariants associated to a matrix A and show how they are related to the Jordan form of A. We obtain also, a very simple criterion to decide when a matrix is diagonalizable

I-Introduction

Let $\mathcal{M}_n(\mathbb{C})$, be the algebra of square matrices of order n, with entries in $\mathbb{C}$ and $\mathrm{GL}_n(\mathbb{C}) = \{P \in \mathcal{M}_n(\mathbb{C}) \mid P \text{ is invertible}\}$.
Let $P \in \mathrm{GL}_n(\mathbb{C})$. We denote by u_P the conjugacy:

$$u_P : \mathcal{M}_n(\mathbb{C}) \longrightarrow \mathcal{M}_n(\mathbb{C})$$
$$A \longmapsto P^{-1}AP$$

We will associate to A, a finite numbers of integers which are invariants under the conjugacy u_P, $\forall P \in \mathrm{GL}_n(\mathbb{C})$ and show how they are related to the Jordan form of A. This study gives also, a very simple finite exact algorithm to determine if a matrix is diagonalizable in $\mathbb{C}$.

Remark

We call "the Jordan form of A":
1) The number and the order of the blocks with distinct eigenvalues.
2) The description of the over diagonal terms of each block $\epsilon_1, \epsilon_2, ..., \epsilon_{r-1}$, in such a way:

$$(\epsilon_1, \epsilon_2, ..., \epsilon_{r-1}) = (E_1, 0, E_2, 0, ..., E_s, N)$$

where:

$E_i = (1, 1, ..., 1)$, $1 \leq i \leq s$, $\mathrm{card}E_{i+1} \leq \mathrm{card}E_i$, $0 \leq i \leq s-1$, $N = (0, 0, ..., 0)$ or $\emptyset$.

II- Numerical invariants associated to A

Let $A \in \mathcal{M}_n(\mathbb{C})$ and f_A be the linear morphism of $\mathcal{M}_n(\mathbb{C})$ defined by:

$$f_A : \mathcal{M}_n(\mathbb{C}) \longrightarrow \mathcal{M}_n(\mathbb{C})$$
$$X \longmapsto AX - XA$$

Let $P \in \mathrm{GL}_n(\mathbb{C})$, u_P be the conjugacy associated to P, and $A' = P^{-1}AP$. We have:

II-1 Proposition

$$f_{A'} = u_P o f_A o u_{P^{-1}} \quad , \quad \forall k \in \mathbb{N} \ .$$

Proof

$$\begin{aligned} f_{A'}(X) &= A'X - XA' = P^{-1}APX - XP^{-1}AP \\ &= P^{-1}[A(PXP^{-1}) - (PXP^{-1})A]P = u_P(f_A(u_{P^{-1}}(X))) \end{aligned}$$

So:

$$f_{A'} = u_P o f_A o u_{P^{-1}} \quad \text{and} \quad f_{A'}^k = u_P o f_A^k o u_{P^{-1}} \quad , \quad \forall k \in \mathbb{N} \ .$$

II-2 Corollary

$\mathrm{rank}(f_A^k)$, $k \in \mathbb{N}$ is invariant under the conjugacy u_P , $\forall P \in \mathrm{GL}_n(\mathbb{C})$.

Proof

Let $A \in \mathcal{M}_n(\mathbb{C})$ and $A' = u_P(A)$. We have:

$$\begin{aligned} X \in \mathrm{Ker} f_{A'}^k &\iff f_{A'}^k(X) = 0 \iff u_P o f_A^k o u_{P^{-1}}(X) = 0 \\ &\iff f_A^k(u_{P^{-1}}(X)) = 0 \iff u_{P^{-1}}(X) \in \mathrm{Ker} f_A^k \\ &\iff X \in u_P(\mathrm{Ker} f_A^k) \end{aligned}$$

So: $\mathrm{Ker} f_{A'}^k = u_P(\mathrm{Ker} f_A^k)$ and $\dim(\mathrm{Ker} f_{A'}^k) = \dim(\mathrm{Ker} f_A^k)$. These integers are invariant under the conjugacy u_P , $\forall P \in \mathrm{GL}_n(\mathbb{C})$, so $\mathrm{rank}(f_A^k)$ too.
Remark that: $\mathrm{rank}(f_A^{k+1}) \leq \mathrm{rank}(f_A^k)$, $\forall k \in \mathbb{N}$. This suite of invariants is finished.

II-3 Theorem

A is diagonalizable if and only if $\mathrm{rank} f_A = \mathrm{rank} f_A^2$.

Proof

We have, always:

$$\begin{aligned} \mathrm{rank} f_A = \mathrm{rank} f_A^2 &\iff \mathrm{Im} f_A = \mathrm{Im} f_A^2 \iff \mathrm{Ker} f_A = \mathrm{Ker} f_A^2 \\ &\iff \mathrm{Ker} f_A \cap \mathrm{Im} f_A = \{0\} \iff \mathcal{M}_n(\mathbb{C}) = \mathrm{Ker} f_A \oplus \mathrm{Im} f_A \quad . \end{aligned}$$

Let us prove : A diagonalizable $\iff \mathrm{Ker} f_A \cap \mathrm{Im} f_A = \{0\}$.
Let $P \in \mathrm{GL}_n(\mathbb{C})$ such that $A' = u_P(A)$ is a Jordan form of A . We have:

$$X \in \mathrm{Ker} f_A \cap \mathrm{Im} f_A \iff X' = u_P(X) \in \mathrm{Ker} f_{A'} \cap \mathrm{Im} f_{A'} \quad .$$

and

$$X' \in \mathrm{Ker} f_{A'} \cap \mathrm{Im} f_{A'} \iff f_{A'}(X') = 0 \ \text{ and } \ \exists Y' \in \mathcal{M}_n(\mathbb{C}) \ \text{ such that } \ X' = f_{A'}(Y') \ .$$

Put: $X' = (x'_{ij})_{1\le i,j\le n}$, $Y' = (y'_{ij})_{1\le i,j\le n}$.
If A is diagonalizable, we have: $A' = \mathrm{diag}(\lambda_1, \lambda_2, ..., \lambda_n)$.
Let $X' \in \mathrm{Ker} f_{A'} \cap \mathrm{Im} f_{A'}$.

$$f_{A'}(X') = 0 \iff A'X' - X'A' = 0 \iff (\lambda_i - \lambda_j)x'_{ij} = 0 \ , \ 1 \le i,j \le n \quad (1)$$
$$X' = f_{A'}(Y') \iff X' = A'Y' - Y'A' \iff (\lambda_i - \lambda_j)y'_{ij} = x'_{ij} \ , \ 1 \le i,j \le n \quad (2)$$

If $\lambda_i - \lambda_j \neq 0$, so by (1): $x'_{ij} = 0$. If $\lambda_i = \lambda_j$, so by (2) : $x'_{ij} = 0$ and $X' = 0$ and $X = 0$.
If A is not diagonalizable A' has at least one Jordan block of the type:

$$J = \begin{pmatrix} \lambda & 1 & \dots & 0 \\ 0 & \lambda & \dots & 0 \\ \vdots & \vdots & \ddots & \vdots \\ 0 & 0 & \dots & 1 \\ 0 & 0 & \dots & \lambda \end{pmatrix}$$

with

$$A' = \left(\begin{array}{c|c|c} & 0 & 0 \\ \hline 0 & J & 0 \\ \hline 0 & 0 & \end{array}\right) \ .$$

The matrix

$$X' = \left(\begin{array}{c|c|c} 0 & 0 & 0 \\ \hline 0 & X'' & 0 \\ \hline 0 & 0 & 0 \end{array}\right)$$

with

$$X'' = \begin{pmatrix} 0 & \dots & 0 & 1 \\ 0 & \dots & 0 & 0 \\ \vdots & \ddots & \vdots & \vdots \\ 0 & \dots & 0 & 0 \end{pmatrix}$$

belongs to $\mathrm{Ker} f_{A'} \cap \mathrm{Im} f_{A'}$ because we can easily verify that:

$$A'X' - X'A' = 0 \ \text{ and } \ X' = A'Y' - Y'A'$$

with

$$Y' = \left(\begin{array}{c|c|c} 0 & 0 & 0 \\ \hline 0 & Y'' & 0 \\ \hline 0 & 0 & 0 \end{array}\right)$$

and

$$Y'' = \begin{pmatrix} 0 & \dots & 0 & 0 \\ 0 & \dots & 0 & 1 \\ \vdots & \ddots & \vdots & \vdots \\ 0 & \dots & 0 & 0 \end{pmatrix} .$$

The theorem is proved.

II-Consequences

We have found a very simple exact algorithm to determine if a matrix is diagonalizable on $\mathbb{C}$.
If we use a classical language (Basic, Pascal, C, Fortran, ...), we get numerical errors and can only say that the considered matrix is near of a diagonalizable matrix.
If we use a software of formal calculus (Mathematica, Maple, ...) and matrices with entries in $\mathbb{N}$, we get exact results (for n not too large, $n \leq 10$).

II-5 Property

A is diagonalizable and has n distinct eigenvalues if and only if $\text{rank}f_A = \text{rank}f_A^2 = n^2 - n$.

Proof

If $(\lambda_i)_{1\leq i\leq n}$ are the eigenvalues of A , $(\lambda_i - \lambda_j)_{1\leq i,j\leq n}$ are the eigenvalues of f_A . So, we have:

$$\text{rank}f_A = n^2 - n \iff \lambda_i \neq \lambda_j \ , \ \forall i,j \in \{1,2,...,n\} \ \text{ with } \ i \neq j \quad .$$

III Invariants and Jordan form of A

Let $A \in \mathcal{M}_n(\mathbb{C})$ and $(\text{rank}f_A^k)_{k\in\mathbb{N}}$ the associated invariants, we have:

III-1 Property

For $n \leq 5$, the finite suite $(\text{rank}f_A^k)_{k\in\mathbb{N}}$, characterizes the Jordan form of A .

Proof

For $n \leq 5$, we can easily study the suite $(\text{rank}f_A^k)_{k\in\mathbb{N}}$ with A under a Jordan form. So, we get tabulars (see the schedule) which describe each possible Jordan form of A and it corresponding suite. For example we have for $n = 4$:
$\text{rank}f_A = 12$, $\text{rank}f_A^2 = 10$, $\text{rank}f_A^3 = 8$, $\text{rank}f_A^4 = 7$, $\text{rank}f_A^5 = 6$ which corresponds to the Jordan form:

$$\begin{pmatrix} \lambda_1 & 1 & 0 & 0 \\ 0 & \lambda_1 & 1 & 0 \\ 0 & 0 & \lambda_1 & 0 \\ 0 & 0 & 0 & \lambda_2 \end{pmatrix} .$$

III-2 Case $n = 6$

We do the same study for $n = 6$ (see the schedule). In this case, all the Jordan forms of A are characterized by the suite $(\text{rank} f_A^k)_{k \in \mathbb{N}}$ except in the two following cases:
a)$\text{rank} f_A = \text{rank} f_A^2 = 24$ where the Jordan form is:
$\text{diag}\{\lambda_1, \lambda_1, \lambda_2, \lambda_2, \lambda_3, \lambda_3\}$ or $\text{diag}\{\lambda_1, \lambda_1, \lambda_1, \lambda_1, \lambda_2, \lambda_3\}$.
b)$\text{rank} f_A = \text{rank} f_A^2 = 18$ where the Jordan form is:
$\text{diag}\{\lambda_1, \lambda_1, \lambda_1, \lambda_2, \lambda_2, \lambda_2\}$ or $\text{diag}\{\lambda_1, \lambda_1, \lambda_1, \lambda_1, \lambda_2, \lambda_3\}$.
In order to remove this ambiguity, we add to our invariants a new family which is:

$$d_l = \text{degree of G.C.D.}(p(\lambda), p^{(l)}(\lambda)) \quad , \quad 1 \leq l \leq n-1 \quad .$$

where $p(\lambda)$ is the characteristic polynomial of A . Note that this polynomial is invariant under the conjugacy u_P , $\forall P \in \text{GL}_n(\mathbb{C})$, and can be obtained exactly from A (more precisely from trace(A^k) , $1 \leq k \leq n$, with the Fadeev-Frame or Newton formulas).
We have in this case:
a)For $\text{diag}\{\lambda_1, \lambda_1, \lambda_2, \lambda_2, \lambda_3, \lambda_3\}$: $d_1 = 3$.
For $\text{diag}\{\lambda_1, \lambda_1, \lambda_1, \lambda_2, \lambda_3, \lambda_4\}$: $d_1 = 2$.
b)For $\text{diag}\{\lambda_1, \lambda_1, \lambda_1, \lambda_2, \lambda_2, \lambda_2\}$: $d_1 = 4$.
For $\text{diag}\{\lambda_1, \lambda_1, \lambda_1, \lambda_1, \lambda_2, \lambda_3\}$: $d_1 = 3$.
The ambiguity is removed.

IV Conclusion

It remains to solve the general case and see if the two suites of invariants characterize the Jordan form of A .

Bibliography

N. GASTINEL : Analyse Numérique Linéaire. Herman (Paris).
G.H. GOLUB and C.F. VAN LOAN : Matrix Computations. North Oxford Academic.
F.R. GANTMACHER : The theory of matrices. Chelsea Publ. Co. (New York).
V.I. ARNOLD : On matrices depending on parameters. English Trans. of Russian Math. Survey 26.
F. CHATELIN : Valeurs propres et vecteurs propres des matrices. Colin (Paris)

B. KLARES
UNIVERSITE DE METZ
U.F.R. "M.I.M."-D.M.I.
U.A. C.N.R.S. 399
Ile du Saulcy
F.57045 METZ Cedex 01
FRANCE

SCHEDULE

1) Case $n = 2$

rank f_A^k			diag		over-diag
2	2		λ_1	λ_2	0
0	0		λ_1	λ_1	0
2	1	0	λ_1	λ_1	1

2) Case $n = 3$

rank f_A^k					diag			over-diag	
6	6				λ_1	λ_2	λ_3	0	0
4	4				λ_1	λ_1	λ_2	0	0
6	5	4			λ_1	λ_1	λ_2	1	0
0	0				λ_1	λ_1	λ_1	1	0
4	1	0			λ_1	λ_1	λ_1	1	0
6	4	2	1	0	λ_1	λ_1	λ_1	1	1

3) Case $n = 4$

rank f_A^k							diag				over-diag		
12	12						λ_1	λ_2	λ_3	λ_4	0	0	0
10	10						λ_1	λ_1	λ_2	λ_3	0	0	0
12	11	10					λ_1	λ_1	λ_2	λ_3	1	0	0
8	8						λ_1	λ_1	λ_2	λ_2	0	0	0
10	9	8					λ_1	λ_1	λ_2	λ_2	1	0	0
12	10	8					λ_1	λ_1	λ_2	λ_2	1	0	1
6	6						λ_1	λ_1	λ_1	λ_2	0	0	0
10	7	6					λ_1	λ_1	λ_1	λ_2	1	0	0
12	10	8	7	6			λ_1	λ_1	λ_1	λ_2	1	1	0
0	0						λ_1	λ_1	λ_1	λ_1	0	0	0
6	1	0					λ_1	λ_1	λ_1	λ_1	1	0	0
10	6	2	1	0			λ_1	λ_1	λ_1	λ_1	1	1	0
8	4	0					λ_1	λ_1	λ_1	λ_1	1	0	1
12	9	6	4	2	1	0	λ_1	λ_1	λ_1	λ_1	1	1	1

4) Case $n = 5$

rank f_A^k									diag					over-diag			
20	20								λ_1	λ_2	λ_3	λ_4	λ_5	0	0	0	0
18	18								λ_1	λ_1	λ_2	λ_3	λ_4	0	0	0	0
20	19	18							λ_1	λ_1	λ_2	λ_3	λ_4	1	0	0	0
16	16								λ_1	λ_1	λ_2	λ_2	λ_3	0	0	0	0
18	17	16							λ_1	λ_1	λ_2	λ_2	λ_3	1	0	0	0
20	18	16							λ_1	λ_1	λ_2	λ_2	λ_3	1	0	1	0
14	14								λ_1	λ_1	λ_1	λ_2	λ_4	0	0	0	0
18	15	14							λ_1	λ_1	λ_1	λ_2	λ_3	1	0	0	0
20	18	16	15	14					λ_1	λ_1	λ_1	λ_2	λ_3	1	1	0	0
12	12								λ_1	λ_1	λ_1	λ_2	λ_2	0	0	0	0
16	13	12							λ_1	λ_1	λ_1	λ_2	λ_2	1	0	0	0
18	16	14	13	12					λ_1	λ_1	λ_1	λ_2	λ_2	1	1	0	0
14	13	12							λ_1	λ_1	λ_1	λ_2	λ_2	0	0	0	1
18	14	12							λ_1	λ_1	λ_1	λ_2	λ_2	1	0	0	1
20	17	14	13	12					λ_1	λ_1	λ_1	λ_2	λ_2	1	1	0	1
8	8								λ_1	λ_1	λ_1	λ_1	λ_2	0	0	0	0
14	9	8							λ_1	λ_1	λ_1	λ_1	λ_2	1	0	0	0
18	14	10	9	8					λ_1	λ_1	λ_1	λ_1	λ_2	1	1	0	0
16	12	8							λ_1	λ_1	λ_1	λ_1	λ_2	1	0	1	0
20	18	14	12	10	9	8			λ_1	λ_1	λ_1	λ_1	λ_2	1	1	1	0
0	0								λ_1	λ_1	λ_1	λ_1	λ_1	0	0	0	0
8	1	0							λ_1	λ_1	λ_1	λ_1	λ_1	1	0	0	0
14	8	2	1	0					λ_1	λ_1	λ_1	λ_1	λ_1	1	1	0	0
12	4	0							λ_1	λ_1	λ_1	λ_1	λ_1	1	0	1	0
18	13	8	4	2	1	0			λ_1	λ_1	λ_1	λ_1	λ_1	1	1	1	0
16	9	4	1	0					λ_1	λ_1	λ_1	λ_1	λ_1	1	1	0	1
20	16	12	9	6	4	2	1	0	λ_1	λ_1	λ_1	λ_1	λ_1	1	1	1	1

4) Case $n = 6$

rank f_A^k											diag						over-diag				
30	30										λ_1	λ_2	λ_3	λ_4	λ_5	λ_6	0	0	0	0	0
28	28										λ_1	λ_1	λ_2	λ_3	λ_4	λ_5	0	0	0	0	0
30	29	28									λ_1	λ_1	λ_2	λ_3	λ_4	λ_5	1	0	0	0	0
26	26										λ_1	λ_1	λ_2	λ_2	λ_3	λ_4	0	0	0	0	0
28	27	26									λ_1	λ_1	λ_2	λ_2	λ_3	λ_4	1	0	0	0	0
30	28	26									λ_1	λ_1	λ_2	λ_2	λ_3	λ_4	1	0	1	0	0
24	24										λ_1	λ_1	λ_2	λ_2	λ_3	λ_3	0	0	0	0	0
26	25	24									λ_1	λ_1	λ_2	λ_2	λ_3	λ_3	1	0	0	0	0
28	26	24									λ_1	λ_1	λ_2	λ_2	λ_3	λ_3	1	0	1	0	0
30	27	24									λ_1	λ_1	λ_2	λ_2	λ_3	λ_3	1	0	1	0	1
24	24										λ_1	λ_1	λ_1	λ_2	λ_3	λ_4	0	0	0	0	0
28	25	24									λ_1	λ_1	λ_1	λ_2	λ_3	λ_4	1	0	0	0	0
30	28	26	25	24							λ_1	λ_1	λ_1	λ_2	λ_3	λ_4	1	1	0	0	0
22	22										λ_1	λ_1	λ_1	λ_2	λ_2	λ_3	0	0	0	0	0
26	23	22									λ_1	λ_1	λ_1	λ_2	λ_2	λ_3	1	0	0	0	0
28	26	24	23	22							λ_1	λ_1	λ_1	λ_2	λ_2	λ_3	1	1	0	0	0
28	24	22									λ_1	λ_1	λ_1	λ_2	λ_2	λ_3	1	0	0	1	0
30	27	24	23	22							λ_1	λ_1	λ_1	λ_2	λ_2	λ_3	1	1	0	1	0
24	23	22									λ_1	λ_1	λ_1	λ_2	λ_2	λ_3	0	0	0	1	0
18	18										λ_1	λ_1	λ_1	λ_2	λ_2	λ_2	0	0	0	0	0
22	19	18									λ_1	λ_1	λ_1	λ_2	λ_2	λ_2	1	0	0	0	0
24	22	20	19	18							λ_1	λ_1	λ_1	λ_2	λ_2	λ_2	1	1	0	0	0
26	20	18									λ_1	λ_1	λ_1	λ_2	λ_2	λ_2	1	0	0	1	0
28	23	20	19	18							λ_1	λ_1	λ_1	λ_2	λ_2	λ_2	1	1	0	1	0
30	26	22	20	18							λ_1	λ_1	λ_1	λ_2	λ_2	λ_2	1	1	0	1	1
18	18										λ_1	λ_1	λ_1	λ_1	λ_2	λ_3	0	0	0	0	0
24	19	18									λ_1	λ_1	λ_1	λ_1	λ_2	λ_3	1	0	0	0	0
28	24	20	19	18							λ_1	λ_1	λ_1	λ_1	λ_2	λ_3	1	1	0	0	0

26	22	18									λ_1	λ_1	λ_1	λ_1	λ_2	λ_3	1	0	1	0	0
30	27	24	22	20	19	18					λ_1	λ_1	λ_1	λ_1	λ_2	λ_3	1	1	1	0	0
16	16										λ_1	λ_1	λ_1	λ_1	λ_2	λ_2	0	0	0	0	0
18	17	16									λ_1	λ_1	λ_1	λ_1	λ_2	λ_2	0	0	0	0	1
22	17	16									λ_1	λ_1	λ_1	λ_1	λ_2	λ_2	1	0	0	0	0
26	22	18	17	16							λ_1	λ_1	λ_1	λ_1	λ_2	λ_2	1	1	1	0	0
24	20	16									λ_1	λ_1	λ_1	λ_1	λ_2	λ_2	1	0	1	0	0
28	25	22	20	18	17	16					λ_1	λ_1	λ_1	λ_1	λ_2	λ_2	1	1	1	0	0
24	18	16									λ_1	λ_1	λ_1	λ_1	λ_2	λ_2	1	0	1	0	1
28	23	18	17	16							λ_1	λ_1	λ_1	λ_1	λ_2	λ_2	1	1	0	0	1
26	21	16									λ_1	λ_1	λ_1	λ_1	λ_2	λ_2	1	0	1	0	1
30	26	22	20	18	17	16					λ_1	λ_1	λ_1	λ_1	λ_2	λ_2	1	1	1	0	1
10	10										λ_1	λ_1	λ_1	λ_1	λ_1	λ_2	0	0	0	0	0
18	11	10									λ_1	λ_1	λ_1	λ_1	λ_1	λ_2	1	0	0	0	0
24	18	12	11	10							λ_1	λ_1	λ_1	λ_1	λ_1	λ_2	1	1	0	0	0
22	14	10									λ_1	λ_1	λ_1	λ_1	λ_1	λ_2	1	0	1	0	0
28	23	18	14	12	11	10					λ_1	λ_1	λ_1	λ_1	λ_1	λ_2	1	1	1	0	0
26	19	14	11	10							λ_1	λ_1	λ_1	λ_1	λ_1	λ_2	1	1	0	1	0
30	26	22	19	16	14	12	11	10			λ_1	λ_1	λ_1	λ_1	λ_1	λ_2	1	1	1	1	0
0	0										λ_1	λ_1	λ_1	λ_1	λ_1	λ_1	0	0	0	0	0
10	1	0									λ_1	λ_1	λ_1	λ_1	λ_1	λ_1	1	0	0	0	0
18	10	2	1	0							λ_1	λ_1	λ_1	λ_1	λ_1	λ_1	1	1	0	0	0
16	4	0									λ_1	λ_1	λ_1	λ_1	λ_1	λ_1	1	0	1	0	0
24	17	10	4	2	1	0					λ_1	λ_1	λ_1	λ_1	λ_1	λ_1	1	1	1	0	0
22	11	4	1	0							λ_1	λ_1	λ_1	λ_1	λ_1	λ_1	1	1	0	1	0
18	9	0									λ_1	λ_1	λ_1	λ_1	λ_1	λ_1	1	0	1	0	1
28	22	16	11	6	4	2	1	0			λ_1	λ_1	λ_1	λ_1	λ_1	λ_1	1	1	1	1	0
26	18	10	6	2	1	0					λ_1	λ_1	λ_1	λ_1	λ_1	λ_1	1	1	1	0	1
24	16	8	4	0							λ_1	λ_1	λ_1	λ_1	λ_1	λ_1	1	1	0	1	1
30	25	20	16	12	9	6	4	2	1	0	λ_1	λ_1	λ_1	λ_1	λ_1	λ_1	0	0	0	0	0

M LUSKIN

Numerical analysis of microstructure for crystals with nonconvex energy density*

Abstract. A nonconvex variational problem is derived from a continuum theory for crystals with symmetry-related variants [3-5, 12-14, 17-21], and the numerical analysis of this problem that we gave in [6, 10] is discussed.

Key words. finite element method, error estimates, numerical approximation, variational problem, nonconvex

AMS(MOS) subject classifications. 65N15,65N30, 35J20, 35J70, 73C60

1. Introduction. Many crystals have an energy density with symmetry-related energy wells. Mathematical models for the equilibrium of such crystals give nonconvex variational problems. Such variational problems may fail to attain a minimum value for any admissible deformation. Also, the deformation gradients of minimizing sequences can have oscillations which do not converge strongly enough to allow the evaluation of the weak limit of nonlinear functions of the deformation gradient (such as the energy density or stress) by the function evaluated at the weak limit of the deformation gradient.

We consider that the macroscopic properties of the crystal are described by the weak limit of thermodynamic functions of the deformation gradient evaluated at deformation gradients for a minimizing sequence. Such a theory of microstructure and microscopic properties for crystals has been extensively developed over the past few years [3-5, 15, 18-21].

In §2, we use the continuum theory for the equilibrium of crystals with symmetry-related variants developed in [3-5, 12-13] to show how some one-dimensional models can be derived by constraining the admissible deformations (this development was inspired by a similar development in [2]). In §3, we discuss the optimal order error estimates for finite element approximations of this one-dimensional model which we have given in [10]. These rigorous results validate the computational results for two and three-dimensional problems which we have reported in [7-9].

*This work was supported in part by the National Science Foundation and AFOSR through grants DMS 835-1080 and DMS 871-8881, the ARO through grants DAAL03-88-K-0110 and DAAL03-89-G-0081, the Cray Research Foundation, and by a grant from the Minnesota Supercomputer Institute.

2. Three-dimensional continuum theory. Let $\Omega \subseteq \mathbb{R}^3$ be the reference configuration for a solid crystal and let $y(x) : \Omega \to \mathbb{R}^3$ be the corresponding deformation. We denote the deformation gradient by $F(x) = \nabla y(x)$ and the right Cauchy-Green strain tensor by $C = F^T F$. We shall assume that the temperature of the crystal is equal to that of its environment, and so we can denote the free energy per unit volume by $\Phi(F)$ with the temperature dependence implicit. The bulk energy for the deformation $y(x)$ is then given by

$$\mathcal{E}(y) = \int_\Omega \Phi(\nabla y(x))\, dx.$$

We denote by M^3 the set of 3×3 matrices and by M^3_+ the set of 3×3 real matrices with positive determinant. We further denote by O^3_+ the set of orthogonal 3×3 real matrices with positive determinant (the space of proper rotations) or

$$O^3_+ = \{\, R \in M^3_+ : R^T R = I \,\}.$$

Finally, we denote the set of positive definite, 3×3 real symmetric matrices by S^3_+ or

$$S^3_+ = \{\, U \in M^3_+ : U = U^T \,\}.$$

We assume that the energy density $\Phi : M^3_+ \to \mathbb{R}$ is frame indifferent so that

$$\Phi(RF) = \Phi(F) \qquad \text{for all } R \in O^3_+,\ F \in M^3_+. \tag{2.1}$$

Now every $F \in M^3_+$ has the polar decomposition

$$F = RU$$

where $R \in O^3_+$ and $U \in S^3_+$. Further, $U = C^{1/2}$ is the right stretch tensor where $C = F^T F$ is the Cauchy-Green strain. It then follows from (2.1) that there exists $\Psi : S^3_+ \to \mathbb{R}$ such that

$$\Phi(F) = \Phi(RU) = \Phi(U) = \Psi(U^2) = \Psi(C).$$

We next assume that the energy density Φ has the finite symmetry group

$$\mathcal{G} = \{\, R_1 = I, R_2, \ldots, R_m \,\} \subset O^3_+$$

so that

$$\Phi(FR_i) = \Phi(F) \qquad \text{for all } R_i \in \mathcal{G},\ F \in M^3_+. \tag{2.2}$$

It then follows from (2.1) and (2.2) that

$$\Psi(R_i^T C R_i) = \Psi(C) \qquad \text{for all } R_i \in \mathcal{G},\ C \in S_+^3.$$

Let $C_1 \in S_+^3$ be the deformation gradient with minimum energy (normalized to be zero) so that

$$\Psi(C) > \Psi(C_1) = 0 \tag{2.3}$$

for all $C \notin \{\, R_i^T C_1 R_i : R_i \in \mathcal{G} \,\}$. We take Ω to be the reference configuration for an austenitic phase (which is the lowest energy phase at higher temperatures) with symmetry group $\mathcal{G}$ which is strictly larger that the symmetry group $\mathcal{H}$ of the martensitic phase. More precisely, we have that

$$\mathcal{H} = \{\, R_i \in \mathcal{G} : R_i^T C_1 R_i = C_1 \,\} \subset \mathcal{G},\ \mathcal{H} \neq \mathcal{G},$$

and

$$\{\, R_i^T C_1 R_i : R_i \in \mathcal{G} \,\} = \{\, C_1, \ldots, C_l \,\}$$

where $l = |\mathcal{G}|/|\mathcal{H}| \geq 2$. Hence, it follows from (2.3) that

$$\Psi(C) > \Psi(C_1) = \cdots = \Psi(C_l) = 0 \tag{2.4}$$

for $C \notin \{C_1, \ldots, C_l\}$. Thus, we see that in this case Φ and Ψ are not convex.

An example for the above is the indium-thallium alloy where $\mathcal{G} = \{R_1, \ldots, R_{24}\}$ is the cubic symmetry group and $C_1 = \operatorname{diag}(1+2\epsilon, 1-\epsilon, 1-\epsilon)$ for $0 < \epsilon < 1$. In this case, $|\mathcal{H}| = 8$ and

$$\{R_i^T C_1 R_i : R_i \in \mathcal{G}\} = \{C_1, C_2, C_3\}$$

where $C_2 = \operatorname{diag}(1-\epsilon, 1+2\epsilon, 1-\epsilon)$ and $C_3 = \operatorname{diag}(1-\epsilon, 1-\epsilon, 1+2\epsilon)$.

For the transformation strains C_i, $i = 1, \ldots, l$, define

$$U_i = C_i^{1/2}.$$

Many crystals have the special property that there exists $R \in O_+^3$, $a \in \mathbb{R}^3$, $a \neq 0$, and $n \in \mathbb{R}^3$, $|n| = 1$, such that [3]

$$RU_1 = U_2 + a \otimes n \tag{2.5}$$

where $a \otimes n \in M_3$ is defined by

$$(a \otimes n)v = (n \cdot v)\, a \qquad \text{for all } v \in \mathbb{R}^3$$

(or more generally, $RU_i = R_j + a \otimes n$). Since

$$RU_1 = \left(I + a \otimes U_2^{-T} n\right) U_2,$$

we have that

$$\det U_1 = (1 + a \cdot U_2^{-T} n) \det U_2.$$

However, $R_i^T U_1 R_i = U_2$ for some $R_i \in \mathcal{G}$, so $\det U_1 = \det U_2$ and

$$a \cdot U_2^{-T} n = 0. \tag{2.6}$$

We shall consider deformations $y(x) : \Omega \to \mathbb{R}^3$ constrained to be of the form

$$y(x) = U_2 x + Y(x \cdot n) a \tag{2.7}$$

for continuous functions $Y(\xi) : \mathbb{R} \to \mathbb{R}$. Then by (2.5)

$$\begin{aligned} \nabla y(x) &= U_2 + Y'(x \cdot n) a \otimes n \\ &= Y'(x \cdot n) RU_1 + (1 - Y'(x \cdot n)) U_2. \end{aligned}$$

We define a one-dimensional energy density by

$$\phi(s) = \Phi(sRU_1 + (1 - s)U_2), \qquad s \in \mathbb{R}. \tag{2.8}$$

(We assume temporarily that Φ is defined on $sRU_1 + (1-s)U_2$ for all $s \in \mathbb{R}$ even though $sRU_1 + (1-s)U_2 \notin M_+^3$ for some $s \in \mathbb{R}$.) It follows from (2.4) that

$$\phi(s) > \phi(0) = \phi(1) = 0 \qquad \text{for } s \in \mathbb{R},\ s \neq 0, 1. \tag{2.9}$$

Note that if we let the martensitic state be the reference configuration with material coordinates $z = U_2 x$, then the deformation (2.7) for $\hat{y}(z) = y(U_2^{-1} z) = y(x)$ has the form

$$\hat{y}(z) = z + Y(z \cdot U_2^{-T} n) a. \tag{2.10}$$

Thus, we see from (2.6) that (2.10) describes a shear in the direction a orthogonal to $U_2^{-1} n$ in the new reference configuration of magnitude Y.

We assume for simplicity of exposition that $n = e_1$ and that $\Omega = (0,1) \times (0,1) \times (0,1)$. Then for $y(x)$ of the form (2.7) we have that

$$\int_\Omega \Phi(\nabla y(x))\, dx = \int_0^1 \phi(Y'(\xi))\, d\xi. \tag{2.11}$$

We note that

$$\int_0^1 \phi(Y'(\xi))\, d\xi = 0$$

for any $Y(\xi)$ such that $Y'(\xi)$ is a characteristic function taking only the values 0 and 1.

We consider the variational problem

$$\inf \int_\Omega \Phi(\nabla y(x))\, dx \tag{2.12}$$

with boundary conditions $y(x) = [U_2 + \bar{s}a \otimes n]\, x$ for $x \in \Gamma \equiv \partial\Omega - \{0,1\} \times (0,1) \times (0,1)$ where $\bar{s}$, $0 < \bar{s} < 1$, is a fixed constant. We can approximate the boundary conditions on $\Gamma \equiv \partial\Omega - \{0,1\} \times (0,1) \times (0,1)$ by the problem

$$\inf \int_\Omega \Phi(\nabla y(x))\, dx + \frac{\kappa}{2} \int_\Gamma |y(x) - [U_2 + \bar{s}a \otimes n]x|^2\, d\sigma \tag{2.13}$$

where $\kappa > 0$ is a penalty parameter.

The problem (2.13) is a model for the mechanical problem of a crystal constrained so that $y(x) \sim [U_2 + \bar{s}a \otimes n]x$ for $x \in \Gamma$ by a restoring traction on the normal plane to Γ which is proportional to the displacement of $y(x)$ from $[U_2 + \bar{s}a \otimes n]x$ on Γ or

$$\frac{\partial \Phi}{\partial F}(\nabla y(x)) \cdot \nu + \kappa\{y(x) - [U_2 + \bar{s}a \otimes n]x\} = 0$$

for $x \in \Gamma$.

We expect that the solution to (2.13) will converge to the solution to (2.12) as $\kappa \to \infty$. For deformations of the form

$$y(x) = U_2 x + Y(x \cdot n)a$$

we have that

$$\begin{aligned}
\int_\Gamma |y(x) - [U_2 + \bar{s}a \otimes n]x|^2\, d\sigma \\
&= \int_\Gamma |Y(x \cdot n)a - \bar{s}[a \otimes n]x|^2\, d\sigma \\
&= 4|a|^2 \int_0^1 |Y(\xi) - \bar{s}\xi|^2\, d\xi.
\end{aligned}$$

Thus, we consider in §3 the equivalent one-dimensional variational problem

$$\inf \int_0^1 \phi(Y'(\xi)) + \hat{\kappa}(Y(\xi) - \bar{s}\xi)^2\, d\xi \tag{2.14}$$

for $0 < \bar{s} < 1$ and $\hat{\kappa} > 0$.

We shall assume that

$$\lambda_1 \min\{s^2, (1-s)^2\} \le \phi(s) \le \lambda_2 s^2 \tag{2.15}$$

for material constants λ_1 and λ_2. The results in §3 show that ϕ need only be defined in an arbitrarily small neighborhood of $s = 0$ and $s = 1$. Thus, the upper bound in (2.15) is unnecessary, and it is not necessary to extend $\Phi(F)$ outside M^3_+ (in particular, $\Psi(C)$ need only be defined in a neighborhood of C_1 and C_2).

3. One-dimensional analysis. In this section, we consider a more general problem than (2.14). We assume that the energy density $\phi(s)$ for the shear deformation of a crystal satisfies

$$\begin{aligned} \lambda_1 \min\{(s-s_L)^2,(s-s_U)^2\} \le \phi(s) \le \lambda_2(|s|+1)^2 \\ \phi(s_L)=\phi(s_U)=0 \end{aligned} \tag{3.1}$$

where λ_1 and λ_2 are material constants and $s_L < s_U$. We model the bulk energy of the crystal by

$$\mathcal{E}(v) = \int_0^1 [\phi(v'(x)) + (v(x)-f(x))^2]\,dx, \tag{3.2}$$

where $v(x)$ is the shear displacement and

$$f(x) = \bar{s}x + b \qquad \text{for } x \in I$$

for $s_L < \bar{s} < s_U$ (the following results also hold for $\hat{\kappa} \neq 1$ in (2.14)).

It is well-known (see Theorem 1 below, for instance) that

$$\inf_{v\in H^1} \mathcal{E}(v) = 0,$$

but that there does not exist $u \in H^1$ such that $\mathcal{E}(u) = 0$. (To see this, assume that $\mathcal{E}(u) = 0$. Since $\int_0^1 (u(x)-f(x))^2\,dx = 0$, we have that $u(x) = f(x)$ a.e. Thus, $\phi(u') = \phi(f') = \phi(\bar{s}) > 0$. So, $\int_0^1 \phi(u'(x))\,dx > 0$. This contradicts the assumption that $\mathcal{E}(u) = 0$.)

We gave an optimal order error analysis in [10] for the minimization of the functional $\mathcal{E}$ over finite element spaces, $\mathcal{M}_h$. To define $\mathcal{M}_h$, let the mesh length $h = 1/M$ for some $M \in \mathbb{N}$; let the vertex points $x_i = ih$ for $i = 0, \ldots, M$; and let the subintervals $I_i = (x_{i-1}, x_i)$ for $i = 1, \ldots, M$. The finite element space $\mathcal{M}_h$ is defined to be the space of piecewise linear, continuous functions

$$\mathcal{M}_h \equiv \{\, v \in C(I) : v|_{I_i} \text{ is linear for } i = 1, \ldots, M \,\}.$$

The approximate solutions $u_h \in \mathcal{M}_h$ satisfy

$$\mathcal{E}(u_h) = \min_{v_h \in \mathcal{M}_h} \mathcal{E}(v_h) \equiv E_h. \tag{3.3}$$

For $\hat{s} \in \mathbb{R}$ and $\alpha > 0$ denote the closed ball of radius α at $\hat{s}$ by

$$\mathcal{B}_\alpha(\hat{s}) = \{\, s : \hat{s} - \alpha \le s \le \hat{s} + \alpha \,\}$$

and denote the closed neighborhood $\mathcal{N}_\alpha$ of $\{s_L, s_U\}$ by

$$\mathcal{N}_\alpha = \mathcal{B}_\alpha(s_L) \cup \mathcal{B}_\alpha(s_U).$$

It is proven in [10] that if $h \le 4\alpha^2\lambda_1/(s_U - s_L)^2$, then

$$u_h'(x) \in \mathcal{N}_\alpha \qquad \text{for } x \in I.$$

Equivalently,

$$u_h'(x) \in \mathcal{N}_{h^{1/2}(s_U - s_L)/2\lambda_1^{1/2}} \qquad \text{for } x \in I.$$

The thermodynamic properties of crystals (stress, energy, pressure) depend nonlinearly on the deformation gradient. We can represent the microscopic properties of the crystal by the microscopic density $F(x, s)$ for the thermodynamic property F. For the deformation u_h the microscopic density is $F(x, u_h'(x))$. Theorem 1 shows that the finite element approximation, $F(x, u_h'(x))$, converges weakly to the correct macroscopic density

$$\gamma F(x, s_L) + (1-\gamma) F(x, s_U).$$

Our main theorem states that the approximate strains, $u_h'(x)$, are locally in the state s_L with probability

$$\gamma = \frac{s_U - \bar{s}}{s_U - s_L} \tag{3.4}$$

and are locally in the state s_U with probability $1 - \gamma = (\bar{s} - s_L)/(s_U - s_L)$. In the jargon of the calculus of variations [4, 5, 24], the measure

$$\nu_x \equiv \gamma \delta_{s_L} + (1-\gamma)\delta_{s_U} \qquad \text{for } 0 < x < 1, \tag{3.5}$$

where $\delta_{\hat{s}}$ is the Dirac delta function with unit mass concentrated at $\hat{s}$ for $\hat{s} = s_L,\ s_U$, is the unique Young measure associated with minimizing (3.2).

We define the Sobolev space $\mathcal{V}$ to be the space of functions $F(x, s) : I \times \mathcal{N}_\alpha \to \mathbb{R}$ such that

$$\frac{\partial F}{\partial s} \in L^2(I, L^\infty(\mathcal{N}_\alpha)) \tag{3.6}$$

and

$$\frac{\partial}{\partial x}\left[F(x, s_U) - F(x, s_L)\right] \in L^1(I)$$

with norm (which depends on $\alpha > 0$)

$$\|F\|_{\mathcal{V}} = \left[\int_0^1 \left|\frac{\partial F}{\partial s}(x, \cdot)\right|^2_{L^\infty(\mathcal{N}_\alpha)} dx\right]^{1/2}$$
$$+ |F(0, s_U) - F(0, s_L)| + \int_0^1 \left|\frac{\partial}{\partial x}\left[F(x, s_U) - F(x, s_L)\right]\right| dx.$$

We proved the following theorem in [10].

THEOREM 1. *For $h \le \min\{4\lambda_1, 4\alpha^2\lambda_1/(s_U - s_L)^2\}$ and $F \in \mathcal{V}$ we have the estimate*

$$\left| \int_0^1 [F(x, u_h'(x)) - \gamma F(x, s_L) - (1-\gamma)F(x, s_U)]\, dx \right| \le \left[\frac{(s_U - s_L)}{2\lambda_1^{1/2}} + \frac{1}{2\lambda_1^{1/2}} + 4 \right] \|F\|_{\mathcal{V}} h. \tag{3.7}$$

To express the rate of convergence of $u_h'(x)$ we define the operator norm on the dual of $\mathcal{V}$, $\mathcal{V}^*$, by

$$\|L\|_{\mathcal{V}^*} = \sup_{F \in \mathcal{V}} \frac{|\langle L, F \rangle|}{\|F\|_{\mathcal{V}}} \tag{3.8}$$

for $L \in \mathcal{V}^*$. For $h \le 4\alpha^2\lambda_1/(s_U - s_L)^2$, we can identify with u_h' the functional $L_{u_h'} \in \mathcal{V}^*$ defined by

$$\langle L_{u_h'}, F \rangle \equiv \int_0^1 F(x, u_h')\, dx$$

and we identify with $\nu \equiv \gamma\delta_{s_L} + (1-\gamma)\delta_{s_U}$ the functional $L_\nu \in \mathcal{V}^*$ defined by

$$\langle L_\nu, F \rangle \equiv \int_0^1 [\gamma F(x, s_L) + (1-\gamma)F(x, s_U)]\, dx.$$

We then have the following theorem which is a direct consequence of (3.7).

THEOREM 2. *For $h \le \min\{4\lambda_1, 4\alpha^2\lambda_1/(s_U - s_L)^2\}$ we have that*

$$\|L_{u_h'} - L_\nu\|_{\mathcal{V}^*} \le \left[\frac{(s_U - s_L)}{2\lambda_1^{1/2}} + \frac{1}{2\lambda_1^{1/2}} + 4 \right] h. \tag{3.9}$$

We also have that

$$\langle L_{u_h'}, F \rangle \le \|F\|_{L^1(I, C(\mathcal{N}_\alpha))} \qquad \text{for } F \in L^1(I, C(\mathcal{N}_\alpha)).$$

Thus, $L_{u_h'}$ is uniformly bounded in the operator norm (with norm 1) in the dual of $L^1(I, C(\mathcal{N}_\alpha))$. Since $\mathcal{V}$ is dense in $L^1(I, C(\mathcal{N}_\alpha))$ the above result implies that

$$\int_0^1 [F(x, u_h'(x)) - \gamma F(x, s_L) - (1-\gamma)F(x, s_U)]\, dx \to 0$$

for all $F \in L^1(I, C(\mathcal{N}_\alpha))$ or $L_{u_h'} \to L_\nu$ as $h \to 0$ in the weak*-topology of

$$L^1(I, C(\mathcal{N}_\alpha))^* = L^\infty(I, C(N_\alpha)^*) = L^\infty(I, M(\mathcal{N}_\alpha))$$

where $M(\mathcal{N}_\alpha)$ is the space of real Borel measures on N_α with bounded variation.

REFERENCES

[1] R. A. ADAMS, *Sobolev Space*, Academic Press, New York, 1975.

[2] J. M. BALL, P. J. HOLMES, R. D. JAMES, R. L. PEGO, P. J. SWART, *On the dynamics of fine structure*, preprint, 1990.

[3] J. M. BALL AND R. D. JAMES, *Fine phase mixtures as minimizers of energy*, Arch. Rational Mech. Anal., 100 (1987), pp. 13–52.

[4] J. M. BALL AND R. D. JAMES, *Experimental tests of a theory of fine microstructure*, preprint, 1990.

[5] M. CHIPOT AND D. KINDERLEHRER, *Equilibrium configurations of crystals*, Arch. Rational Mech. Anal, 103 (1988), pp. 237–277.

[6] C. COLLINS, D. KINDERLEHRER, AND M. LUSKIN, *Numerical approximation of the solution of a variational problem with a double well potential*, IMA preprint 605, 1990, SIAM J. Numer. Anal. (to appear).

[7] C. COLLINS AND M. LUSKIN, *The computation of the austenitic-martensitic phase transition*, in *Partial Differential Equations and Continuum Models of Phase Transitions, Lecture Notes in Physics 344, M. Rascle, D.Serre, and M. Slemrod, eds.*, Springer-Verlag, Berlin, New York, 1989, pp. 34–50.

[8] C. COLLINS AND M. LUSKIN, *Computational results for phase transitions in shape memory materials*, in *Smart Materials, Structures, and Mathematical Issues, C. Rogers, ed.*, Technomic Publishing Co., Lancaster, Pennsylvania, 1989, pp. 198–215.

[9] C. COLLINS AND M. LUSKIN, *Numerical modeling of the microstructure of crystals with symmetry-related variants*, in *Proceedings of the ARO US-Japan Workshop on Smart/Intelligent Materials and Systems*, Technomic Publishing Co., Lancaster, Pennsylvania, to appear.

[10] C. COLLINS AND M. LUSKIN, *Optimal order error estimates for the finite element approximation of the solution of a nonconvex variational problem*, Math. Comp. (submitted).

[11] R. E. EDWARDS, *Functional Analysis*, Holt, Rinehart, and Winston, Inc., New York, 1965.

[12] J. L. ERICKSEN, *Some constrained elastic crystals*, in *Material Instabilities in Continuum Mechanics and Related Problems, J. M. Ball, ed.*, Oxford University Press, Oxford, 1987, pp. 119–137.

[13] J. L. ERICKSEN, *Constitutive theory for some constrained elastic crystals*, Int. J. Solids and Structures, 22 (1986), pp. 951–964.

[14] I. FONSECA, *Variational methods for elastic crystals*, Arch. Rat. Mech. Anal., 97 (1985), pp. 189–220.

[15] I. FONSECA, *The lower quasiconvex envelope of the stored energy function for an elastic crystal*, J. Math. Pures et Appl., 67 (1988), pp. 175–195.

[16] D. A. FRENCH, *On the convergence of finite element approximations of a relaxed variational problem*, SIAM J. Numer. Anal. (1990), pp. 419–436.

[17] R. JAMES, *Basic principles for the improvement of shape-memory and related materials*, in *Smart Materials, Structures, and Mathematical Issues, C. Rogers, ed.*, Technomic Publishing Co., Lancaster, PA, 1989.

[18] R. D. JAMES, *Microstructure and weak convergence*, in *Material Instabilities in Continuum Mechanics and Related Problems, J. M. Ball, ed.*, Oxford University Press, Oxford, 1987, pp. 175–196.

[19] R. D. JAMES AND D. KINDERLEHRER, *Theory of diffusionless phase transitions*, in *Partial Differential Equations and Continuum Models of Phase Transitions, Lecture Notes in Physics 344, M. Rascle, D.Serre, and M. Slemrod, eds.*, Springer-Verlag, Berlin, New York, 1989, pp. 51–84.

[20] D. KINDERLEHRER, *Remarks about equilibrium configurations of crystals*, in *Material Instabilities in Continuum Mechanics and Related Problems, J. M. Ball, ed.*, Oxford University Press, Oxford, 1987, pp. 217–242.

[21] R. KOHN, *The relationship between linear and nonlinear variational models of coherent phase transitions*, in *Proceedings of the Seventh Army Conference on Applied Mathematics and Computing, West Point, F. Dressel, ed.*, 1989.

[22] W. RUDIN, *Functional analysis*, McGraw-Hill, New York, 1973.

[23] W. RUDIN, *Real & Complex Analysis*, Third edition, McGraw-Hill, New York, 1987.

[24] L. Tartar, *Compensated compactness and applications to partial differential equations*, in *Nonlinear Analysis and Mechanics: Heriot-Watt Symposium IV, Research Notes 39, R. J. Knops, ed.*, Pitman, London, 1978, pp. 136–212.

Applied Mathematics, California Institute of Technology, Pasadena, California 91125.

J C MIELLOU[1], L GIRAUD[2], A LAOUAR[3] AND P SPITERI[2]

Subdomain decomposition methods with overlapping and asynchronous iterations

In the present paper we study and experiment subdomain decomposition methods in a framework which associates general fixed point iterations on a product space (namely asynchronous iterations), and other numerical methods such as preconditionned conjugate gradient algorithms.

We refer the reader interested in asynchronous iterations to

D. BERTZEKAS and J.N. TSITSIKLIS [1]

EL TARAZI [6], [7] P. SPITERI [23]

J.-C. MIELLOU and P. SPITERI [21]

J.-C. MIELLOU, M. BOULBRACHENE and P. CORTEY-DUMONT [19]

For previous results about asynchronous Schwarz alternating method we refer to

J.-C. MIELLOU G.R. PERRIN and P. SPITERI [20] and also P.L. LIONS (in [3], [12]) which considers iterations with delays (which are one of the aspects of asynchronous iterations). Of course the Proceedings of the first, second (and third) symposia on domain decomposition methods [3], [12] contain a lot of papers about domain decomposition methods (with or without overlapping). In the special case of Schwarz alternating procedure we can also refer to other recent papers such as :

P.E. BJORSTAD and O.B. WIDLUND. [2]

D.J. EVANS et alt [8], [9] KANG LI-SHAN and D.J. EVANS [14] DAUTRAY-LIONS (P. BENILAN) [4] chap. II § 7 about the content of the present article :

---

(1) Laboratoire de Calcul Scientifique et U.A. - C.N.R.S. n° 741 - 16, route de Gray, 25030 BESANÇON CEDEX

(2) U.R.A. - C.N.R.S. n° 1399 - E.N.S.E.E.I.H.T. - 2 rue Charles Camichel - 31071 TOULOUSE CEDEX

(3) Institut de Mathématiques - Université de Annaba - B.P. 12 - El Hadjat - Annaba - Algérie

In section 1 : For elliptic Dirichlet problems we formulate a subdomain decomposition method with overlapping for multiple subdomains in terms of a system of integral equations with subboundary unknowns, the solution of which is approximated by asynchronous relaxation methods. Such a formalism cover both :

- The standart sequential GAUSS-SEIDEL iteration which, associated to a decomposition into two subdomains with overlapping, leads to the classical Schwarz alternating method.

- more sophisticated algorithms including delays associated to communications between parallel processes, in the context of multiple subdomains decomposition.

In section 2 : We study the case of a bidimensionnal Poisson's problem solver for a rectangular domain with a hole that we subdivide into four subdomains. We verify that the general methodology considered in section I, can be applied there and we give an explicit block matricial form for the discretized fixed point problem.

In section 3 : We describe brevily a tridimensionnal Poisson's solver using the bidimensionnal Poisson's solver as a preconditionner for either a block diagonally preconditionned conjugate gradient squared iterative algorithm, or a block S.S.O.R. preconditionned conjugate gradient method.

The experimental results presented are issued from an implementation of these methods on the I.B.M. 3090-400 multiprocessor of the Centre National Universitaire Sud and supported by the centre de Calcul Numérique Intensif (C.N.U.S.C.-I.B.M.). Our study has also been partially supported by the D.R.E.T. contract n° 85/1194.

Some points of view and results have been adapted, in section 1, from the thesis of A. LAOUAR [15] chap. 4, and, in section 3, from the Thesis of L. GIRAUD [10].

Section 1

A subdomain decomposition method with overlapping for elliptic problems :

Let us consider a domain Ω , and subdomains Ω_i $i \in \{1, ..., m\}$ such that :

$$(1\text{-}1) \quad \begin{cases} \Omega \subset R^n \,;\, \Omega_i \subset R^n \quad i \in \{1, \ldots, m\} \\ \text{and :} \\ \Omega = \bigcup_i \Omega_i \end{cases}$$

We note :

$\Gamma_i = \partial \Omega_i \cap \partial \Omega \,;\, \Gamma_i^{'} = \partial \Omega_i \cap \Omega$

We assume more over that :

$$(1\text{-}2) \quad \begin{cases} \text{For any } i \in \{1, ..., m\} \text{ their exists :} \\ \text{- a boundary subset } \gamma_i \subset \bigcup_{\substack{j \in \{1,\ldots,m\} \\ j \neq i}} \partial \Omega_j \\ \text{- an index subset } I(i) \subset \{1,\ldots,m\} \\ \text{such that :} \\ \gamma_i \subset \Omega_i \text{ and } \overline{\gamma_i} \cap \Gamma_i^{'} = \emptyset \text{ (where } \overline{\gamma_i} \text{ is the closing of } \gamma_i) \\ \text{and} \\ \Gamma_i^{'} \subset (\bigcup_{j \in I(i)} \gamma_j), \text{ and more precisely} \\ \Gamma_i^{'} = \bigcup_{j \in I(i)} \gamma_{j,i} \text{ with } \gamma_{j,i} \subset \gamma_j \end{cases}$$

The elliptic problem :

(1-3) $\begin{cases} \text{A is a second order operator with uniformly elliptic} \\ \text{principal part on } \Omega \text{ , and regular coefficients} \end{cases}$

Let us consider given functions $f \in C(\Omega)$ and $g \in C(\partial\Omega)$, to which we associate the Dirichlet problem :

(1-4) $$Au = f_{/\Omega} \, ; \, u_{/\partial\Omega} = g$$

(1-5) $\begin{cases} \text{and we assume that (1-4) has a unique} \\ \text{solution } u \in C(\Omega) \cap H^1(\Omega) \end{cases}$

(1-6) $\begin{cases} \text{we make the same assumptions as above, with} \\ \Omega_i \, , \, \partial\Omega_i \; f_i = f_{/\Omega_i} \; g_i \in C(\partial\Omega_i), \text{ for the problems :} \\ Au_i = f_{i/\Omega_i} \, ; \, u_{i/\partial\Omega_i} = g_i \end{cases}$

A system of coupled elliptic equations :

(1-7) $\begin{cases} \text{For any } i \in \{1,\ldots,m\} \\ Au_i = f_i = f/_{\Omega_i} \\ u_{i/\Gamma_i} = g_i = g/_{\Gamma_i} \, ; \, u_{i/\gamma_{j,i}} = u_{j/\gamma_{j,i}} \; \forall \, j \in J(i) \end{cases}$

We have the following evident :

Proposition 1 : u being the solution of (1-4), then $u_i = u/_{\Omega_i}$ is solution of the system (1-7).

Another formulation for the system (1-7) :

Let us introduce :

a) The uncoupled subproblems :

$$
(1\text{-}8) \quad \begin{cases} A\overline{w}_i = f_i = f/_{\Omega_i} \\ \overline{w}_{i/\Gamma_i} = g_i = g/_{\Gamma_i} \; ; \; w_{i/\Gamma'_i} = 0 \end{cases}
$$

We write :

$$
\Psi_i = \overline{w}_{i/\gamma_i}
$$

b) Let us consider on Ω_i , the linear integral operator with Poisson's kernel :

$\mathcal{T}_i$, which to a given function Φ_i , defined on Γ'_i , associates $\mathcal{T}_i(\Phi_i)$ defined by :

$$
Aw_i = 0/_{\Omega_i} \; ; \; w_{i/\Gamma_i} = 0 \; ; \; w_{i/\Gamma'_i} = \Phi_i
$$

and then :

$$
\mathcal{T}_i(\Phi_i) = w_{i/\gamma_i}
$$

Taking into account assumption (1-2), let us now consider functions φ_j , defined on γ_j , and consider :

$\tilde{\Phi}_i$ defined on Γ'_i by $\tilde{\Phi}_{i/\gamma_{j,i}} = \varphi_{j/\gamma_{j,i}}$

$$
(1\text{-}9) \quad \begin{cases} \text{Then, } j \in I(i) \;\; T_i\,(\ldots,\varphi_j,\ldots) \\ \text{is defined by } \mathcal{T}_i(\tilde{\Phi}_i) \\ \text{and let } \; T(\varphi) = \{\ldots,\, T_i\,(\ldots,\varphi_j,\ldots),\ldots\}^t \end{cases}
$$

Let us consider now the fixed point system of boundary integral equations :

$$
(1\text{-}10) \qquad \overset{*}{\varphi}_i = T_i\,(\ldots,\overset{*}{\varphi}_j,\ldots) + \Psi_i \; ; \text{ or } \; \varphi^* = T(\varphi^*) + \Psi
$$

c) assuming that we can get

φ_i^*, $i \in \{1,\dots,m\}$ solution of (1-10).

Let us consider the uncoupled subproblems

$$\text{(1-11)} \qquad Au_i = f_{i/\Omega_i}\ ;\ u_{/\Gamma_i^i} = g_i\ ;\ u_{/\Gamma_i^{i\,\prime}} = \varphi_i$$

<u>Proposition 2</u> : With the assumptions and notations (1-1) to (1-3), (1-5), (1-6) let us assume the uniqueness of the solution of problem (1-10) in the space $\prod_{i=1}^{n} C(\vec{\gamma}_i)$, then the solution u_i, $i \in \{1,\dots,m\}$ of the uncoupled subproblems (1-11), is the unique solution of the system (1-7) in $\prod_i (C(\Omega_i) \cap H^1(\Omega_i))$

and u solution of (1-4) satisfies

$$u_{/\Omega_i} = u_i$$

<u>Proof</u> : $\varphi_i^* = u_{/\gamma_i}$ $i \in \{1,\dots,m\}$ is clearly solution of (1-10), but :

$\varphi_i^{**} = u_{i/\gamma_i}$ is also solution of (1-10).

By our uniqueness assumption : $\varphi_i^{**} = \varphi_i^*$, $i \in \{1,\dots,m\}$.

and the uniqueness of the solution of each elliptic subproblems (1-11) achieves the proof.

<u>General iterative fixed point methods : the asynchronous iterations :</u>

In order to approximate the solution of (1-10), let us consider the following iterative algorithm :

$$
(1\text{-}12) \quad \begin{cases} \varphi^0 = \{\ldots,\varphi_i^0,\ldots\} \in \prod_{j=1}^{m} C(\gamma_j) \text{ is an initial data, and we define the} \\ \text{sequence } \{\varphi^p\} \text{ satisfying :} \\ \varphi_i^{p+1} = T_i\,(\ldots,\varphi_j^{s_j(p)},\ldots) + \Psi_i \ \forall\, i \in J(p), \text{ and } \varphi_i^p,\ \forall\, i \notin J(p) \end{cases}
$$

where :

$J = \{J(p)\}$ is a sequence of non void subsets of $\{1,2,\ldots,m\}$ ($J(p)$ is the subset of the indexes of the components of φ which are chooses to be updated at the pth iteration).

$S = \{s_1(p),\ldots,s_j(p),\ldots,s_m(p)\}_{p\in N}$ is a sequence of elements of N^m.

Moreover J and S satisfy :

(1-13) $\quad \forall\, j \in \{1,2,\ldots,m\}$ card $\{p \in N, i \in J(p)\}$ is infinite.

(1-14) $\quad \forall\, j \in \{1,2,\ldots,m\}\ \forall\, j \in N,\ s_j(p) \le p$.

(1-15) $\quad \forall\, j \in \{1,2,\ldots,m\} \lim_{p\to\infty} s_j(p) = +\infty$

Remark 1 : a relaxation parameter can be introduced by replacing the affine mapping :

$$\varphi \to T\varphi + \Psi$$

by :

$$\varphi \to \omega\,(T\varphi + \Psi) + (1-\omega)\,\varphi$$

where $\omega \in\,]0,2[$

Remark 2 : In the general framework considered here above we have to deal with discontinous boundary data, the corresponding theoretical results shall be presented elsewhere (J.-C. MIELLOU [18]) but in the present paper we consider here after a simpler situation.

Section 2
A bidimensionnal Poisson's solver

In the $\{x_1,x_2\}$ plane let us consider $\Omega =]-1,+1[\times]-1,+1[\ominus [-a_1, a_1] \times [-a_2,a_2]$ where the given numbers a_1, a_2 are such that $0 < a_i < 1$.

So $[-a_1,a_1] \times [-a_2,a_2]$ is a hole of Ω .

Let us consider the straight line

$S = \{\{x_1,x_2\}/x_1 = 1\}$, and then :

$\hat{\Gamma}_2 = \partial\,\Omega_{\wedge}\,S\,;\,\hat{\Gamma}_1 = \partial\,\Omega \ominus \hat{\Gamma}_2$

Let $f \in C(\Omega)$, and $y \in C(\hat{\Gamma}_1)$ we have to solve :

$$(2\text{-}1) \quad \begin{cases} -\Delta\ u + qu = f/_{\Omega} & (q \geq 0) \\ u/_{\hat{\Gamma}_1} = y\ ;\ \dfrac{\partial u}{\partial n}|_{\hat{\Gamma}_2} = 0 \end{cases}$$

<u>Remark</u> : In problem (2-1) appears a mixt boundary condition which we did not considered in section I, but by a symetry with respect to S we can extend problem (2-1) to a larger (symetric with respect to S) domain ; wich becomes a pure Dirichlet problem. All the theoretical difficulties related to the NEUMAN condition on $\hat{\Gamma}_2$ can be overcomed here by such an argument, valid in the case of problem (2-1).

The subdomain decomposition

(2-2) $\quad$ h is a positive discretization parameter such that : $h < \min\{a_1,a_2\}$

We first define

$$(2\text{-}3)\quad \begin{cases} \Omega_I = \{\{\{x_1,x_2\} \,/\, x_1 \geq -h \text{ and } x_2 \geq -h\} \\ \cup\ \{\{x_1,x_2\} \,/\, x_1 \leq h \text{ and } x_2 \leq h\}\}_{\wedge}\ \Omega \\ \Omega_{II} = \{\{\{x_1,x_2\} \,/\, x_1 \geq -h \text{ and } x_2 \leq h\} \\ \cup\ \{\{x_1,x_2\} \,/\, x_1 \leq h \text{ and } x_2 \geq -h\}\}_{\wedge}\ \Omega \end{cases}$$

The subdomains Ω_I and Ω_{II} are not connexe, so let us consider then :

The decomposition of Ω_I :

$$(2\text{-}4)\quad \begin{cases} \Omega_I = \Omega_1 \cup \Omega_3 \\ \Omega_1 = \{\{x_1,x_2\} \,/\, x_1 \geq -h \text{ and } x_2 \geq -h\}_{\wedge}\ \Omega \\ \Omega_3 = \{\{x_1,x_2\} \,/\, x_1 \leq h \text{ and } x_2 \leq h\}_{\wedge}\ \Omega \end{cases}$$

The decomposition of Ω_{II} :

$$(2\text{-}5)\quad \begin{cases} \Omega_{II} = \Omega_2 \cup \Omega_4 \\ \Omega_2 = \{\{x_1,x_2\} \,/\, x_1 \geq -h \text{ and } x_2 \leq h\}_{\wedge}\ \Omega \\ \Omega_4 = \{\{x_1,x_2\} \,/\, x_1 \leq h \text{ and } x_2 \geq -h\}_{\wedge}\ \Omega \end{cases}$$

We are now able to take into account the notations and assumptions of section I.

$$(2\text{-}6)\qquad \Omega = \bigcup_{i=1,2,3,4} \Omega_i$$

is the analogue of (1-1).

And if we note :

$$
(2\text{-}7)\quad \begin{cases} \gamma_{i,j} = \partial\Omega_j \cap \Omega_i \text{ for } i,j = 1,2,3,4 \text{ and } i \neq j \\ \text{we get that with the notations of (1-2)} \\ \gamma_1 = \gamma_{1,2} \cup \gamma_{1,4}; \gamma_2 = \gamma_{2,1} \cup \gamma_{2,3}; \\ \gamma_3 = \gamma_{3,2} \cup \gamma_{3,4}; \gamma_4 = \gamma_{4,1} \cup \gamma_{4,3}; \\ \text{where the subset indexes } I(i) \text{ are :} \\ I(1) = \{2,4\} \ ; \ I(2) = \{1,3\} \ ; \ I(3) = \{3,4\} \ ; \ I(4) = \{3,1\} \end{cases}
$$

By a simple use of the maximum principle we get then the following :

<u>Proposition 3</u> : The linear mapping T associated to the specific decomposition (2-6), (2-7), by (1-9) is a strict contraction on $\prod_{i=1}^{4} C(\bar{\gamma}_i)$, which can also be considered as

$\prod_{\substack{i,j=1 \\ i\neq j}}^{4} C(\gamma_{i,j})$ with the uniform norm

$$\max_{\substack{i,j=1,2,3,4 \\ i\neq j}} |\varphi_{i,j}|_\infty$$

<u>Corollary 1</u> : The asynchronous iterations (1-12) applied to the affine mapping $T(\varphi) + \Psi$ associated to (2-6), (2-7) converges in uniform norm towards $\varphi^* \in \prod_{i=1}^{4} C(\bar{\gamma}_i)$ which is the restriction of the unique solution u of the problem (2-1) to the boundary subset $\prod_{i=1}^{4} \gamma_i$.

<u>Proof</u> : Results form the use of <u>Proposition 3</u> and of El TARAZI's theorem [6], [7].

<u>Discrete problem</u> : After a suitable P_1 finite element discretization, and a suitable preprocessing, we get a finite dimensionnal linear problem, which can be viewed as a fixed point problem associated to a finite dimensionnal affine mapping, the matrix of which is yet named T . Then we have to distinguish among the submatrices of T :

1) submatrices $T_{i,j}$ which perform strong interactions between neighbouring boundary subsets.

2) submatrices $t_{i,j}$ which perform weak interactions between boundary subsets which are far from each other.

We have written here after the linear matricial problem with two orderings :

The first one is such that if we use then the standart GAUSS-SEIDEL's method, we get the standart Schwarz alternating method applied to the initial decomposition of Ω :

$$\Omega = \Omega_I \cup \Omega_{II}$$

The second one corresponds to associate closely in the same diagonal blocks the strong interactions associated to neighbouring boundary subsets $\gamma_{i,j}$.

In both case the behaviour of the iterative algorithms is highly delay tolerant with respect to the updating of weak interactions : $t_{i,j}\ \varphi_{j,i}^{s_j(p)}$ with $(s_j(p) \leq p)$. This point can be used in order to :

- avoid unefficient computations (G. LEBLANC et J.-C. MIELLOU [17]) even in a purely sequential approach.

- allow a great freedom in communication policies, in the case of parallel processes.

These two matricial forms of our problem are :

$$\left|\begin{array}{cccc|cccc}
I_{1,2} & & & & -T_{1,2} & -t_{1,4} & 0 & 0 \\
& I_{1,4} & & & -t_{1,2} & -T_{1,4} & 0 & 0 \\
& & I_{3,2} & & 0 & 0 & -T_{3,2} & -t_{3,4} \\
& & & I_{3,4} & 0 & 0 & -t_{3,2} & -T_{3,4} \\
\hline
-T_{2,1} & 0 & -t_{2,3} & 0 & I_{2,1} & & & \\
0 & -T_{4,1} & 0 & -t_{4,3} & & I_{4,1} & & \\
-t_{2,1} & 0 & -T_{2,3} & 0 & & & I_{2,3} & \\
0 & -t_{4,1} & 0 & -T_{4,3} & & & & I_{4,3}
\end{array}\right\| \left|\begin{array}{c}
\varphi_{1,2} \\ \varphi_{1,4} \\ \varphi_{3,2} \\ \varphi_{3,4} \\ \varphi_{2,1} \\ \varphi_{4,1} \\ \varphi_{2,3} \\ \varphi_{4,3}
\end{array}\right| \left|\begin{array}{c}
\Psi_{1,2} \\ \Psi_{1,4} \\ \Psi_{3,2} \\ \Psi_{3,4} \\ \Psi_{2,1} \\ \Psi_{4,1} \\ \Psi_{2,3} \\ \Psi_{4,3}
\end{array}\right|$$

$$\begin{array}{cccccccc}
I_{1,2} & -T_{1,2} & 0 & -t_{1,4} & 0 & 0 & 0 & 0 \\
-T_{2,1} & I_{2,1} & 0 & 0 & -t_{2,3} & 0 & 0 & 0 \\
0 & -t_{1,2} & I_{1,4} & -T_{1,4} & 0 & 0 & 0 & 0 \\
0 & 0 & -T_{4,1} & I_{4,1} & 0 & 0 & -t_{4,3} & 0 \\
0 & 0 & 0 & 0 & I_{3,2} & -T_{3,2} & 0 & -t_{3,4} \\
-t_{2,1} & 0 & 0 & 0 & -T_{2,3} & I_{2,3} & 0 & 0 \\
 & 0 & & 0 & 0 & -t_{3,2} & I_{3,4} & -T_{3,4} \\
0 & -t_{4,1} & 0 & 0 & 0 & 0 & -T_{4,3} & I_{4,3}
\end{array}
\left\| \begin{array}{c}
\varphi_{1,2} \\ \varphi_{2,1} \\ \varphi_{1,4} \\ \varphi_{4,1} \\ \varphi_{3,2} \\ \varphi_{2,3} \\ \varphi_{3,4} \\ \varphi_{4,3}
\end{array} \right|
\quad
\left| \begin{array}{c}
\Psi_{1,2} \\ \Psi_{2,1} \\ \Psi_{1,4} \\ \Psi_{4,1} \\ \Psi_{3,2} \\ \Psi_{2,3} \\ \Psi_{3,4} \\ \Psi_{4,3}
\end{array} \right|$$

The different steps of the bidimensionnal Poisson's solver are :

The matrices $T_{i,j}$, $t_{i,j}$ being known, at each use of the bidimensionnal Poisson's solver we have to :

A) solve the uncoupled subproblems (1-8) and get the function Ψ_i on γ_i (and so $\Psi_{i,j}$ on $\gamma_{i,j}$).

B) to solve the discrete fixed point analogue of (1-10) (the just above linear problem), by the iterations (1-12).

C) solve the uncoupled subproblems (1-11) with $\varphi_i = \varphi_i^*$ obtained in step B .

In the steps A) and C) various methods have been used :

- Diagonally preconditionned (MANTEUFEL)

- Conjugate gradient squared noted C.G.S.

- S.S.O.R. Preconditionned conjugate gradient noted S.S.O.R.

- Multigrid. noted M.G.

N.B. steps A) and C) are respectively called Pre and Post processing in the tables of results.

Section 3
A three-dimensionnal solver

We have considered a discretization by planes of three-dimensionnal Poisson's problem, leading to a system of bidimensionnal elliptic equation :

$$\begin{pmatrix} -k^2\Delta_2+2I & -I & 0 & \\ -I & & & -I \\ & 0 & -I & -k^2\Delta_2+2I \end{pmatrix} \begin{pmatrix} u_1 \\ u_i \\ u_N \end{pmatrix} = \begin{pmatrix} f_1 \\ f_i \\ f_N \end{pmatrix}$$

where k is a discretization parameter for the third dimension x_3 and :

$$\Delta_2 = \frac{\partial^2}{\partial x_1^2} + \frac{\partial^2}{\partial x_2^2}$$

We have used then the bidimensionnal Poisson's problem solver in order to get :

- a diagonally preconditionned conjugate gradient squared iteration algorithm

- a S.S.O.R. preconditionned conjugate gradient method.

We don't describe these well known methods (cf. and give now P. LASCAUX et R. THEODOR [16], P. SONNENVELD et alt. [22] experimental results in which for the S.S.O.R. preconditionned conjugate gradient we use either a lexicographical ordering or a red black ordering.

For the resolution of system (1-10) in all cases the delays associated to computation of weak interaction have been used, but from a computer science point of view, we have used either synchronous or completely asynchronous parallel processes all theses situations are quoted in our results :

VECTORIAL PERFORMANCES (Times)

3D Solver	Mode	2D Solver with pre and post-processing: C.G.S.	S.S.O.R.	MG
S.S.O.R. lexico	scal.	119.48	95.87	63.23
	vect.	68.20	50.56	30.24
C.G.S.	scal.	////////////////////////	////////////////////////	82.71
	vect.	////////////////////////	////////////////////////	37.02
S.S.O.R. Red - Black	scal.	////////////////////////	////////////////////////	86.48
	vect.	////////////////////////	////////////////////////	41.35

260 000 unknowns

S.S.O.R. lexico	scal.	////////////////////////	8.18	6.58
	vect.	////////////////////////	4.33	4.49

16 000 unknowns

PARALLEL PERFORMANCES (Times)

		2D Solver with pre and post-processing		
3D Solver	Mode	C.G.S.	S.S.O.R.	MG
S.S.O.R. lexico	sync.	37.62	30.93	17.41
	async.	38.20	28.15	17.43
C.G.S.	sync.	//////////////////////////	//////////////////////////	21.36
	async.	//////////////////////////	//////////////////////////	21.10
S.S.O.R. Red - Black	sync.	//////////////////////////	//////////////////////////	23.82
	async.	//////////////////////////	//////////////////////////	23.53

260 000 unknowns

S.S.O.R. lexico	sync.	//////////////////////////	2.75	2.84
	async.	//////////////////////////	2.37	2.46

16 000 unknowns

PARALLEL PERFORMANCES

3D Solver	Mode	2D Solver with pre and post-processing: C.G.S.	S.S.O.R.	MG
S.S.O.R. lexico	sync.	0.91	0.82	0.87
	async.	0.89	0.90	0.87
C.G.S.	sync.	////////////////////////////	////////////////////////////	0.87
	async.	////////////////////////////	////////////////////////////	0.88
S.S.O.R. R.N.	sync.	////////////////////////////	////////////////////////////	0.87
	async.	////////////////////////////	////////////////////////////	0.88

260 000 unknowns

S.S.O.R. lexico	sync.	////////////////////////////	0.79	0.79
	async.	////////////////////////////	0.91	0.91

16 000 unknowns

Conclusion : The model of subdomain decomposition described in section 1 provides us with a large amount of possibilities of such decompositions and we have described a particular case of them in section 2 for a two dimensionnal geometry. Our experimental implementation for a 3D-problem associates various usual iterative methods to the asynchronous iteration applied to this decomposition.

For other iterative methods for the matricial problems associated to (1-10), see A. LAOUAR [15].

[1] D.P. BERTZEKAS and J.-N. TSITSIKLIS, Parallel and Distributed Computation - Numerical methods, PRENTICE HALL - 1989.

[2] P.E. BJORSTAD and U.B. WIDLUND, To overlap or not to overlap : a note on a domain decomposition method for elliptic problems, SIAM J. SCI. STAT. COMPUT. Vol. 10, n° 5, pp. 1053-61, 1989.

[3] T.F. CHAN, R. GLOWINSKI, J. PERIAUX, O.B. WIDLUND. ed. DOMAIN DECOMPOSITION METHODS, Proc. of the second int. symp. on domain decomposition methods - Los Angeles - 1988, S.I.A.M.

[4] DAUTRAY-LIONS, Analyse mathématique et calcul numérique pour les sciences et les techniques. L'opérateur de Laplace. Vol. 2. Masson - 1987.

[5] DINH Q.U., GLOWINSKI R., MANTEL B., DERIAUX J., PERRIER P. : Subdomain solutions of non linear problems in fluid dipronic on parallel processors. 5ème Col. int. sur les méthodes de calcul scientifique et technique, 16-18 Déc. 81, Versailles.

[6] M.N. EL TARAZI, Contraction et ordre partiel pour l'étude d'algorithmes synchrones et asynchrones en analyse numérique, Thèse - Besançon - 1981.

[7] M.N. EL TARAZI, Some convergence results for asynchronous iterations, numer. Math., 39, 1982, p. 325-340 (article en français).

[8] D.J. EVANS, SHAO JIAN-PING KANG, LI-SHAN and CHEN YU-PING

- The convergence rate of the Schwarz alternating procedure I : For one-dimensionnal probems. Intern. J. Computer Math. 1986. Vol. 20, pp. 157-170.

- The convergence rate of the Schwarz alternating procedure (II) : For two-dimensionnal problems. Intern. J. Computer Math. 1986. Vol. 20, pp. 325-339.

[9] D.J. EVANS, KANG LI-SHAN, CHEN YU-PING and SHAO JIAN-PING, The convergence rate of the Schwarz alternating procedure (IV) - with pseudo boundary relaxation factor. Intern. J. Computer Math. 1987. Vol. 21, pp. 185-203.

[10] L. GIRAUD, Thèse ENSEEIGHT, à paraître.

[11] L. GIRAUD, N. BORGES, J.-C. MIELLOU, P. SPITERI, Méthode de sous-domaine pour la résolution d'un problème de Poisson tridimensionnel, 22ème Congrès National d'Analyse numérique, 28 mai - 1er juin 1990.

[12] R. GLOWINSKI, G.H. GOLUB, G.A. MEURANT, J. PERIAUX, Proc. of the first int. symp. on DOMAIN DECOMPOSITION METHODS FOR PARTIAL DIFFERENTIAL EQUATIONS, S.I.A.M., 1988.

[13] S. GRISOUARD, J.-C. MIELLOU, et P. SPITERI, Résolution du problème de Poisson 3D par une méthode mixte : algorithme C.G.S. préconditionné / méthode alternée de Schwarz parallèle asynchrone. Annexe 5 au rapport Commande sur mémoire n° 85 (1194) DRET.

[14] KANG LI-SHAN and D.-J. EVANS, The convergence rate of the Schwarz alternating procedure (III) - for Reumann problems, Int. J. Comput. Math. 1987. Vol. 21, pp. 85-108.

[15] A. LAOUAR, Aspects de l'analyse numérique de méthodes itératives de point fixe : erreurs d'arrondi, accélération de convergence, sous-domaines. Thèse - Université de Franche-Comté - Besançon. 1988.

[16] P. LASCAUX et R. THEODOR, Analyse numérique matricielle appliquée à l'art de l'Ingénieur - Tome 2, Masson - 1987.

[17] G. LEBLANC et J.-C. MIELLOU, Méthode de Schwarz asynchrone sur le super calculateur CRAY-1. Rapport C^2 VR - 1986.

[18] J.-C. MIELLOU, Subdomain decomposition methods with overlapping for elliptic problems (à paraître)

[19] J.-C. MIELLOU, M. BOULBRACHENE and P. CORTEY-DUMONT, Perturbation of fixed point iterative methods, Advances in parallel processing (à paraître), J.A.I. Presse (U.K.).

[20] J.-C. MIELLOU, G.R. PERRIN and P. SPITERI, An inexpensive method of performance evaluation for subdomain decomposition parallel algorithms of three-dimensionnal elliptical problems, The Journal of Systems and Software. Vol. 6, number 1 & 2, May 1986, pp. 169-173.

[21] J.-C. MIELLOU et P. SPITERI, Un critère de convergence pour des méthodes générales de point fixe, M^2 A.N., Vol. 19, n° 4, 1985, p. 645 à 669.

[22] P. SONNENVELD, P. WESSELING, P.M. de ZEEW, Multigrid and conjugate gradient methods as convergence acceleration techniques - short course on multigrid methods. BRISTOL (1983).

[23] P. SPITERI, Contribution à l'étude de grands systèmes non linéaires, comportement d'algorithmes itératifs, stabilité de systèmes continus. Thèse de Doctorat es Sciences, Besançon, 1984.

R RANNACHER

Defect correction techniques in the finite element method

Institut für Angewandte Mathematik
Universität Heidelberg
Im Neuenheimer Feld 293
D-6900 Heidelberg

Summary. This article surveys some recent theoretical and practical results on the use of defect correction techniques in the finite element method for solving elliptic and parabolic differential equations. Like the classical Richardson h-extrapolation the so-called defect correction "by higher order differences" is an approved method for increasing the accuracy of low order discretization schemes. Its theoretical justification is found in asymptotic error expansions in terms of powers of the mesh size. The traditional proof of such expansions in the context of finite difference schemes relies on discrete maximum principles which severely restricts its range of applicability. Recently a more flexible approach has been developed within the analysis of the finite element Galerkin method which particularly applies to many of the problems arising in structural and fluid mechanics.

0. Introduction

The principle of "defect correction" is one of the very basic tools in numerical analysis. Its classical application is in the context of the solution of linear or nonlinear algebraic systems. In fact, most of the iterative solution methods (even the multigrid approach) may be viewed as applications of the so-called "iterated defect correction". Let a continuous (linear) differential problem $Lu = f$ be discretized resulting in an algebraic equation $L_h u_h = r_h f$, where r_h is some restriction operator mapping continuous functions into grid functions. Then, the often difficult to invert operator L_h is preconditioned by a nearby operator $\tilde{L}_h$ which has a simpler structure. This leads to an iterative process of the form

$$(0.1) \qquad \tilde{L}_h u_h^{(m+1)} = \tilde{L}_h u_h^{(m)} - \left(L_h u_h^{(m)} - r_h f\right),$$

where $L_h u_h^{(m)} - r_h f$ is the "defect" of the iterate $u_h^{(m)}$. Clearly, the convergence of this method depends on the contractivity of the mapping $I-\tilde{L}_h^{-1}L_h$, as there holds

$$\|u_h^{(m)}-u_h\| \le \|I-\tilde{L}_h^{-1}L_h\|^m \|u_h^{(0)}-u_h\| , \quad m \ge 1 . \tag{0.2}$$

Another use of defect correction is in the construction of highly accurate discretizations, which frequently conflicts with the demand for good stability and solution properties. To by-pass this problem, one first chooses a simple lower order scheme $L_h u_h = f_h$ as a base approximation. Then, having computed u_h , one solves the corrected equation

$$L_h\bar{u}_h = L_h u_h - \{\tilde{L}_h u_h - r_h f\} , \tag{0.3}$$

with some higher order approximation $\tilde{L}_h$ to L . Iterating this process one obtains

$$\|u_h^{(m)}-r_h u\| \le q_h^m \|u_h-r_h u\| + \frac{1}{1-q_h} \|L_h^{-1}\| \|\tilde{L}_h r_h u - r_h f\| , \quad m \ge 1 , \tag{0.4}$$

where $q_h \equiv \|I-\tilde{L}_h^{-1}L_h\| < 1$. If $q_h << 1$, only a few correction steps are performed until the full accuracy of the higher order approximation $\tilde{L}_h$ is achieved. However, normally the theoretical justification of this approach is not based on the contraction property but rather on the existence of an asymptotic error expansion for the base scheme of the form

$$u_h = r_h u + h^p r_h e(u) + o(h^p) . \tag{0.5}$$

Here, the coefficient $e(u)$ is independent of h , and the remainder term is bounded in terms of some higher order norm of u . In this case, there holds

$$L_h(\bar{u}_h-r_h u) = (L_h-\tilde{L}_h)(h^p r_h e(u)+o(h^p)) - (\tilde{L}_h r_h - r_h L)u = o(h^p) , \tag{0.6}$$

from which one infers that

$$\bar{u}_h = r_h u + o(h^p) , \tag{0.7}$$

provided that the base scheme is stable, $\|L_h^{-1}\| \le c$. The main advantage of the just described method consists in its low computational complexity as it requires only the inversion of the base approximation L_h . Furthermore, for some applications it is important that the higher order approximation $\tilde{L}_h$ itself does not need to be stable, for $h \to 0$. Below, we will discuss the theoretical justification of this defect correction as well as its application for the solution of elliptic and parabolic differential problems by the finite element Galerkin method. For a general analysis of this and related methods with particular emphasis on ordinary differential equations, we refer to the references [28], [20], [27], [13], and to the survey article [5].

The asymptotic error expansion (0.5) is the central tool in the analysis of the defect correction by "higher order differences". The classical proofs of error expansions for finite difference schemes are based on discrete maximum principles for the base operator L_h , which essentially restricts this approach to second order scalar problems (see [28] and [21]). This limitation can be overcome in the context of the finite element method. Here, the presence of error expansions can be rigorously proven for situations in which the usual maximum principles do not hold, e.g., for the systems and the higher order equations in structural and fluid mechanics. Pioneering work in this direction was done by Q.Lin and his collaborators in [14] and [15]. Their approach has then been developed further in [2] and [22], in order to cover a wider range of problems. A survey of these results for the Richardson h-extrapolation in the finite element method is given in [23]. The theoretical and practical aspects of the closely related defect correction methods are thoroughly investigated in [1].

1. The Defect Correction by "Higher Order Differences"

For simplicity the following discussion is restricted to the usual elliptic model problem

$$-\Delta u = f, \text{ in } \Omega, \quad u_{|\partial\Omega} = 0, \tag{1.1}$$

where $\Omega \subset \mathbf{R}^2$ is a convex polygonal domain. Extensions to more complicated situations including variable coefficients and curved boundaries, fourth-order problems, second order systems and even saddle point problems are possible; see [1], and also [17], [22], [24].

Below, we use the notation $L^p(\Omega)$, $W^{m,p}(\Omega)$, $W_0^{m,p}(\Omega)$ for the usual Lebesgue and Sobolev function spaces on Ω, and $\|\cdot\|_p = \|\cdot\|_{p;\Omega}$ and $\|\cdot\|_{m,p} = \|\cdot\|_{m,p;\Omega}$ for the corresponding norms. The inner product and norm of $L^2(\Omega)$ are simply denoted by $(\cdot,\cdot)$ and $\|\cdot\|$, respectively. $C_0(\Omega)$ is the space of all continuous functions on Ω with zero boundary values. The symbol c is used for a generic constant varying with the context, but which is always independent of the discretization parameter h, and of the solution u.

Let $\mathbf{T}_h = \{T\}$ be finite triangulations of Ω, where the parameter h represents the mesh size, i.e., the maximum diameter of the triangles in $\mathbf{T}_h$. The family $\{\mathbf{T}_h\}_{h>0}$ is assumed to be uniformly regular in the usual sense (see [6]), i.e., to satisfy the uniform shape and size condition. The corresponding spaces of "linear" finite elements are

$$S_h = \{v_h \in C_0(\Omega) : v_{h|T} \in P_1(T), \forall T \in \mathbf{T}_h\} \subset W_0^{1,2}(\Omega).$$

The natural nodal interpolation operator and the L^2-projection onto S_h are denoted by I_h and P_h, respectively. Further, we introduce a "discrete" Laplacian operator Δ_h: $S_h \rightarrow S_h$ through the relation

$$(-\Delta_h v_h, \varphi_h) = (\nabla v_h, \nabla\varphi_h), \quad \forall \varphi_h \in S_h. \tag{1.2}$$

With this notation, the Ritz projection $u_h = R_h u \in S_h$ of the solution u of (1.1) is defined by

$$-\Delta_h u_h = P_h f. \tag{1.3}$$

For this scheme we have the quasi-optimal (local) L^∞-error estimate (see [10], and also [6])

$$(1.4) \qquad \|u-u_h\|_{\infty;\Omega''} \le c_\varepsilon h^{2-\varepsilon}\left\{\|u\|_{2,\infty;\Omega'}+\|u\|_{2,2}\right\},$$

for any fixed subdomains $\Omega''\subset\subset\Omega'\subset\subset\Omega$, and for any $\varepsilon>0$. Here and below, the ε-dependence could be traded for a logarithm $|\ln(h)|^\gamma$. The estimate (1.4) holds for general uniformly regular meshes. On certain uniform meshes ("piecewise three-directional") it can be extended to an asymptotic error expansion of the form

$$(1.5) \qquad u_h = I_h u + h^2 I_h e(u) + h^4 \rho_h(u),$$

where $e(u)$ is a fixed function and the remainder term $\rho_h(u)\in S_h$ satisfies

$$(1.6) \qquad \|\rho_h(u)\|_{\infty;\Omega''} \le c_\varepsilon(u)h^{-\varepsilon}.$$

The proof of this result is based on sharp L^1-estimates for discrete fundamental solutions which are derived by a Hilbert space method using weighted L^2-norms; for details we refer to [10], [2], [23], and to the literature cited therein.

In many applications the accuracy of the second order discretization (1.3) does not suffice. However, the use of higher degree finite elements may be prohibited, since the corresponding system matrix is expensive to generate and hard to invert. In addition, for the low order schemes, often especially adapted efficient solvers (e.g., multigrid) are available. In this case the classical Richardson h-extrapolation or, alternatively, the defect correction approach provide ways for increasing the accuracy while retaining the solution economy of the base scheme. To outline this procedure, we start with its very simplest version related to *linear* finite elements.

Let $\widetilde{\mathbf{T}}_h = \mathbf{T}_{2h}$ be a fixed triangulation and let a refined triangulation $\mathbf{T}_h$ be constructed from $\widetilde{\mathbf{T}}_h$ by subdividing each element into four congruent subtriangles. Then, in view of the expansion (1.5), the Richardson extrapolation constructs an improved approximation $\bar{u}_{2h}\in S_{2h}$ by combining the two solutions $u_{2h}\in S_{2h}$ and $u_h\in S_h$,

$$(1.7) \qquad \bar{u}_{2h} \equiv \tfrac{1}{3}\{4I_{2h}u_h - u_{2h}\} = I_{2h}u + O(h^{4-\varepsilon}).$$

This approach requires the computation of the discrete solution u_h on the two meshes $\mathbf{T}_{2h}$ and $\mathbf{T}_h$, and directly yields an improved approximation only on the coarser mesh $\mathbf{T}_{2h}$. These shortcomings are avoided in the defect correction method which works as follows. On the "macro-triangulation" $\widetilde{\mathbf{T}}_h$ we introduce the space $\widetilde{S}_h$ of quadratic finite elements,

$$\widetilde{S}_h = \{\tilde{v}_h\in C_0(\Omega) : \tilde{v}_{h|\widetilde{T}}\in P_2(\widetilde{T}),\ \forall \widetilde{T}\in\widetilde{\mathbf{T}}_h\} \subset W_0^{1,2}(\Omega),$$

and denote by $\tilde{I}_h: C_0(\Omega) \to \widetilde{S}_h$ the corresponding interpolation operator (with the vertices of $\mathbf{T}_h$ as nodal points). The approximation in $\widetilde{S}_h$ is of third order, $\|v-\tilde{I}_h v\|_\infty \le ch^3\|v\|_{3,\infty}$.

Having computed the approximation $u_h \in S_h$ from (1.3), the solution of the corrected equation

$$(1.8) \qquad -\Delta_h \bar{u}_h = P_h f + \left\{ \Delta_h R_h \tilde{I}_h u_h + P_h f \right\},$$

is taken as a new approximation to u. The analysis of this method is based on the error expansion (1.5) and on the L^∞-stability of the Ritz projection R_h,

$$(1.9) \qquad \|R_h v\|_\infty \le c_\varepsilon h^{-\varepsilon} |||\Delta_h R_h v|||_{-2,\infty;h} .$$

Here, $|||\cdot|||_{-2,\infty;h}$ is a certain discrete Sobolev norm satisfying

$$(1.10) \qquad |||\Delta_h R_h v|||_{-2,\infty;h} \le c\|v\|_\infty$$

Observing that $\tilde{I}_h = \tilde{I}_h I_h$, we obtain

$$-\Delta_h(\bar{u}_h - I_h u) = -\Delta_h R_h (1-\tilde{I}_h)(u_h - I_h u) + \Delta_h R_h (\tilde{I}_h u - u) .$$

and, in virtue of (1.9) and (1.10),

$$(1.11) \qquad \|\bar{u}_h - I_h u\|_\infty \le c_\varepsilon h^{-\varepsilon} \left\{ \|(1-\tilde{I}_h)(u_h - I_h u)\|_\infty + \|u - \tilde{I}_h u\|_\infty \right\} .$$

The expansion (1.6) implies that the base error $u_h - I_h u$ is "smooth",

$$(1.12) \qquad (1-\tilde{I}_h)(u_h - I_h u) = h^2 (1-\tilde{I}_h) e(u) + h^4 (1-\tilde{I}_h) \rho_h(u) .$$

Hence, using the L^∞-stability of the operator $\tilde{I}_h$, $\|\tilde{I}_h v\|_\infty \le c\|v\|_\infty$, we find that

$$(1.13) \qquad \|\bar{u}_h - I_h u\|_\infty = O(h^{3-\varepsilon}) .$$

By a refined analysis using a super approximation property of quadratic finite elements on uniform meshes (see [18]), the order of (local) convergence of this scheme can be even further improved to $O(h^{4-\varepsilon})$. This effect is confirmed by the results shown in Table 1.1, below, but it is somewhat delicate as it breaks down if the mesh $\mathbf{T}_h$ is only piecewise uniform. The procedure (1.8) may be extended to an "iterated" defect correction using higher degree polynomials in the defect computation, which, however, requires an even higher degree of regularity for the solution u ; see [9] for some numerical test calculations. Still this approach may be appropriate on only piecewise uniform meshes in the neighborhood of the interior macro-vertices; see Table 2.1 in the next section.

The model problem (1.1), with the solution $u(x,y) = 16x(1-x)y(1-y)$, is solved on a uniform mesh $\mathbf{T}_h$ and on a piecewise uniform mesh $\mathbf{T}_h^\times$, as shown in Fig. 1.2 below. The following table contains the L^∞-errors and the corresponding orders of convergence m ,

(a) for the base scheme (1.3), and after one step of defect correction on the mesh $\mathbf{T}_h$,
(b) after one step of defect correction on the mesh $\mathbf{T}_h^\times$, measured over Ω , and then also measured over a subdomain $\Omega' \subset \Omega$ having a distance 1/8 from the center point.

Table 1.1. Defect correction by quadratic elements (from [11])

h	$\|e_h^{(a)}\|_\infty$	m	$\|\bar{e}_h^{(a)}\|_\infty$	m	$\|\bar{e}_h^{(b)}\|_\infty$	m	$\|\bar{e}_h^{(b)}\|_{\infty;\Omega'}$	m
2^{-3}	.741 (-2)		.169 (-2)		.290 (-2)		.101 (-2)	
2^{-4}	.195 (-2)	1.98	.110 (-3)	3.95	.610 (-3)	2.25	.130 (-3)	2.95
2^{-5}	.491 (-3)	1.99	.693 (-5)	3.99	.144 (-3)	2.08	.166 (-4)	2.98

Obviously, the scheme (1.9) is sensitive with respect to mesh distortions. A very accurate and robust method is obtained through using cubic elements in the defect calculation. For this, there are essentially two options as indicated in Fig. 1.2, below: the Lagrange version (L) and the Hermite version (H) (or, of course, a mixture of both on several parts of the mesh).

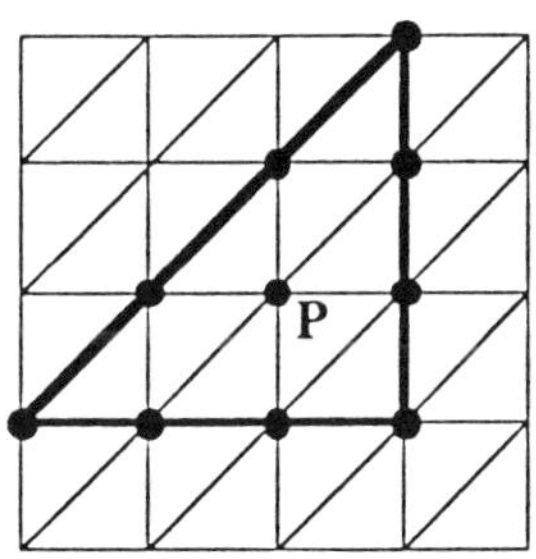

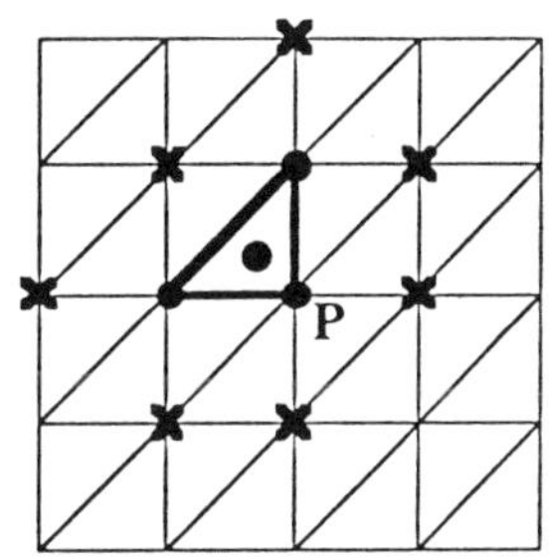

Fig. 1.1. Two versions of cubic interpolation.

In version (L), cubic polynomials are constructed on macro-triangles, each consisting of 9 triangles of the original triangulation $\mathbf{T}_h$, by interpolating the nodal values $u_h(P)$. The corresponding interpolation operator $\tilde{I}_h^{(L)}$ satisfies $\tilde{I}_h^{(L)} I_h = \tilde{I}_h^{(L)}$ and

$$(1.15) \qquad \|v - \tilde{I}_h^{(L)} v\|_\infty \leq ch^4 \|v\|_{4,\infty} .$$

Hence, a variant of the above analysis yields the result

$$(1.16) \qquad \bar{u}_h = I_h u + O(h^{4-\varepsilon}) .$$

In the Hermite case (H) the required derivative nodal values $\nabla u(P)$ are not defined for functions in S_h. As a substitute one uses central difference quotients along the edges of $\mathbf{T}_h$, as indicated in the above picture, and higher order one-sided differences along $\partial\Omega$, and along the macro-edges. Again, the corresponding interpolation operator $\tilde{I}_h^{(H)}$ satisfies $\tilde{I}_h^{(H)} I_h = \tilde{I}_h^{(H)}$, but the analogue of the error estimate (1.15) does not hold with the order $O(h^4)$, unless the interior degree of freedom is chosen in a very special way. However, it is proven in [1] that

$$(1.16) \qquad \|R_h(v-I_h^{(H)}v)\|_\infty \le ch^{4-\varepsilon}\|v\|_{4,\infty},$$

which suffices for the above argument..

The following table contains some results for the model problem (1.1), with the exact solution $u(x,y) = 16x(1-x)\text{cox}(\pi x)y(1-y)\cos(\pi y)$, obtained

(a) by directly using the cubic Hermite element in the base scheme (1.3),
(b) by Richardson h-extrapolation of the base scheme,
(c) by defect correction with the cubic Hermite interpolation on the uniform mesh $\mathbf{T}_h$,
(d) by the same method as in (c), but on the piecewise uniform mesh $\mathbf{T}_h^\times$.

Table 1.2. Several schemes based on cubic interpolation (from [1])

h	$\|e_h^{(a)}\|_\infty$	m	$\|\bar{e}_h^{(b)}\|_\infty$	m	$\|\bar{e}_h^{(c)}\|_\infty$	m	$\|\bar{e}_h^{(d)}\|_{\infty;\Omega'}$	m
2^{-3}	.245 (-3)		.155 (-3)		.568(-2)		.552(-4)	
2^{-4}	.322 (-4)	2.93	.109 (-4)	3.82	.566(-3)	3.32	.430(-5)	3.68
2^{-5}	.253 (-5)	3.67	.705 (-6)	3.95	.414(-4)	3.77	.284(-6)	3.92

We note that a further defect correction step does not much reduce the size of $\|\bar{e}_h^{(c)}\|_\infty$. The defect correction of (isoparametric) bi-linear finite elements on a uniform quadrilateral mesh by bi-cubic Hermite interpolation gives results which are competetive with those obtained by the direct use of the cubic Hermite element in the base scheme (1.3). This superiority of the bi-cubic over the cubic interpolation is documented by the error plots shown below.

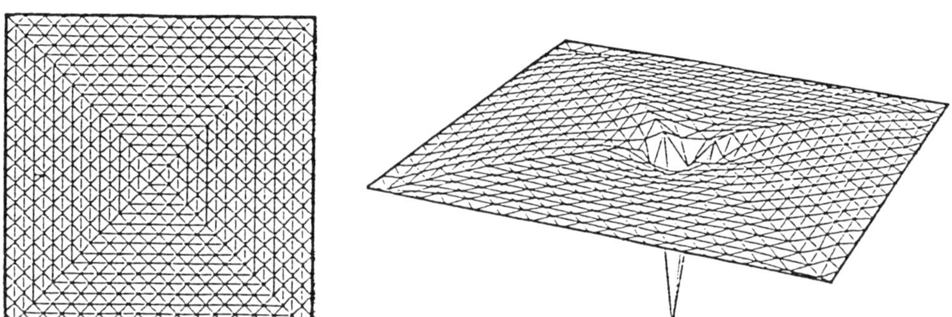

Fig. 1.2. Cubic defect correction (× 500, scaling factor); from [1].

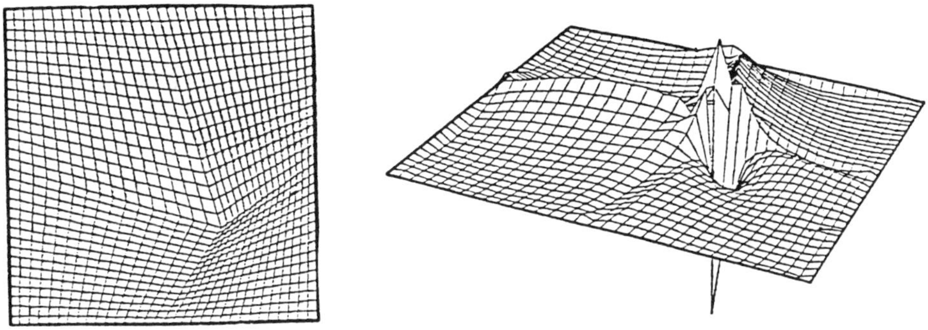

Fig. 1.3. Bi-cubic defect correction (× 40000); from [1].

We list the properties which a higher order interpolation operator $\tilde{I}_h$ should possess in order to fit into the described framework of defect correction:

(i) $\tilde{I}_h: C_0(\Omega) \rightarrow C_0(\Omega)$, (ii) $\tilde{I}_h I_h = \tilde{I}_h$, (iii) $\|R_h(\tilde{I}_h v - v)\|_\infty = O(h^{4-\varepsilon})$.

It is not difficult to construct correction processes of even higher order, say $O(h^6)$, by using quintic Lagrange or Hermite interpolation. This, however, is of only minor practical value because of the limitations for the regularity of the solution u on polygonal domains.

The described defect correction approach has several advantages over the method which directly uses higher order elements in the base scheme (1.3):

- The system matrices have only 7 nonzero entries in contrast, e.g., to the 21 of the cubic Hermite element.
- The base system (1.3) can be very efficiently solved by multigrid methods.
- The availability of two solutions u_h and $\bar{u}_h$ of different orders allows an estimate of the discretization error, $\bar{u}_h - I_h u = (u_h - \bar{u}_h)/3 + O(h^{4-\varepsilon})$.
- The correction operators $\tilde{L}_h$ are not required to be stable, for $h \rightarrow 0$. In fact, they do not even need to constitute a proper approximation scheme for problem (1.1).

The last property becomes important in the approximation of problems with algebraic constraints as, e.g., the incompressibility condition in the Navier-Stokes problem; for a detailed discussion of this aspect we refer to [1]. The obvious major disadvantage of the defect correction method is its lower flexibility compared to the traditional higher order elements, since its success depends on the uniformity of the underlying mesh which is a problem in the case of complicated domains and curved boundaries. This aspect will be further discussed in the next section. Finally, we remark that, of course, also eigenvalue problems may be treated by extrapolation and defect correction techniques (see [18], [4], [19], and [1] for a thorough analysis). Corresponding results for nonlinear problems are rather rare, yet. However, since the L^∞-error analysis for nonlinear finite element schemes is very much developed (see [7], and the literature cited therein) results in this direction will appear in the near future.

2. More General Domains

In this section we discuss some extensions of the defect correction approach described above in order to cope with real life situations including complex regions in two and three dimensions, curved boundaries and reentrant corners or edges. In order to handle more complicated domains Ω, with a partially curved boundary, one uses "parametric" elements. First, a macro-decomposition $\mathbf{T}_H$ of Ω, of width $H \approx 1$, is chosen, such that each macro-element can be smoothly mapped onto a fixed reference element $\hat{T}$. Then, any uniform triangulation of $\hat{T}$ induces smoothly distorted uniform decompositions of each of the macro-elements and thereby a "piecewise" uniform triangulation $\mathbf{T}_h$ of Ω. On piecewise uniform meshes error expansions like (1.5) remain valid (see [17] and [2]),

(2.1) $$u_h = I_h u + h^2 I_h e(u) + h^4 \rho_h(u) .$$

In this case the coefficient $e(u)$ contains additional terms which become singular as the point x approaches an interior macro-vertex P of $\mathbf{T}_H$,

(2.2) $$e(u;x) \approx \ln(|x-P|) D^2 u(P) , \quad \text{as } x \to P .$$

It is remarkable that the expansion (2.1) is of fourth order even across the macro-edges where the mesh is not fully uniform. Since the stability estimate (1.12) holds for general uniformly regular triangulations, the above results on the defect correction schemes carry over to piecewise uniform triangulations, at least in some distance to the macro-vertices (see [2] and [1]),

(2.3) $$\bar{u}_h = I_h u + O(h^{4-\varepsilon}) .$$

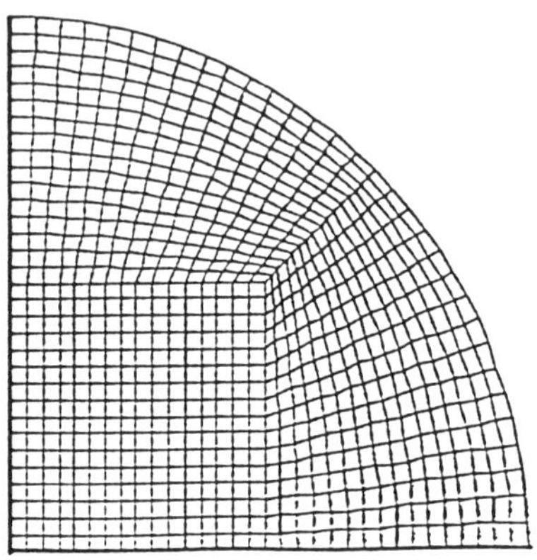

Fig. 2.1. A piecewise uniform mesh of a non-polygonal domain.

The following error plots show the result of the Richardson h-extrapolation for the approximation of the model problem (1.1) by bi-linear elements (from [1]).

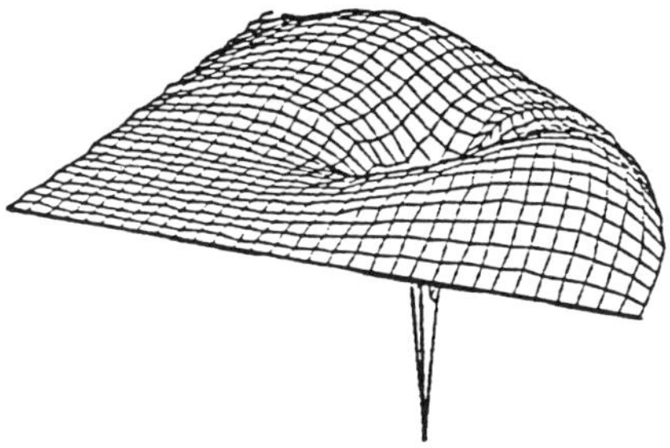

Fig. 2.2. Bi-linear elements (× 3000).

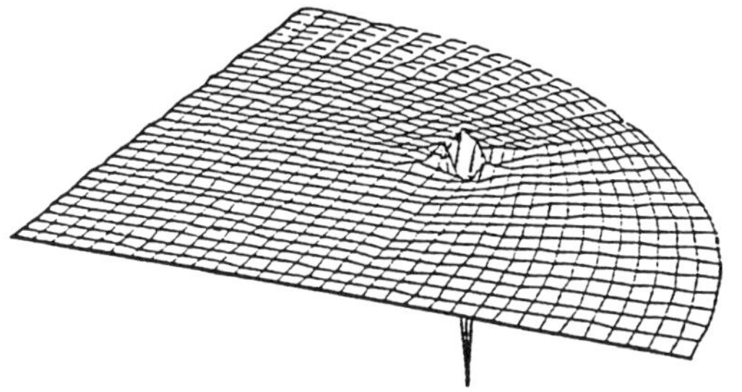

Fig. 2.3. Richardson h-extrapolation (× 30000).

In order to obtain a high accuracy uniformly over Ω , one has to iterate the defect correction process. In virtue of the contraction property of the iteration operator, a good approximation is normaly achieved after only 2-3 steps. Since, on the subdomain Ω' , the fourth order accuracy is already guaranteed for the first iterate $\bar{u}_h$, the further correction steps need to be performed only on a small neighborhood of the macro-vertices. Table 2.1, below, shows the results of the iterated defect correction (with cubic Hermite interpolation) for problem (1.1) on the mesh $T_h^{\times}$.

Table 2.1. Iterated defect correction on a piecewise uniform mesh (from [1]).

h	$\|e_h\|_\infty$	$\|\bar{e}_h^{(1)}\|_\infty$	$\|\bar{e}_h^{(2)}\|_\infty$	$\|\bar{e}_h^{(3)}\|_\infty$	$\|\bar{e}_h^{(4)}\|_\infty$	$\|\bar{e}_h^{(5)}\|_\infty$
2^{-3}	.152 (-1)	.299 (-2)	.710 (-3)	.144 (-3)	.165 (-3)	.635 (-4)
2^{-4}	.498 (-2)	.719 (-3)	.191 (-3)	.588 (-3)	.214 (-4)	.104 (-4)
2^{-5}	.153 (-2)	.176 (-3)	.465 (-4)	.141 (-4)	.486 (-5)	.217 (-5)

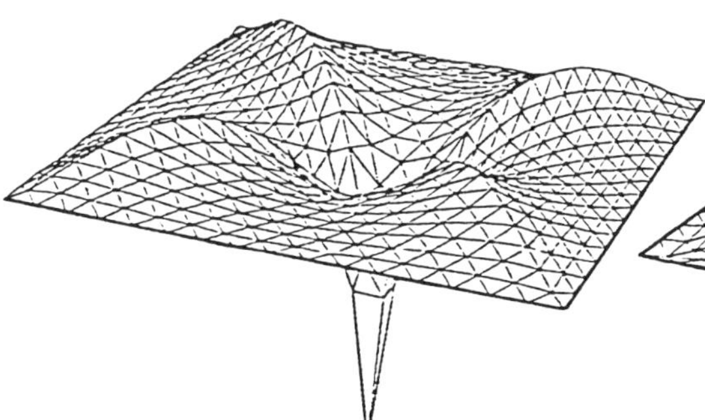

Fig. 2.4. Linear elements (× 100).

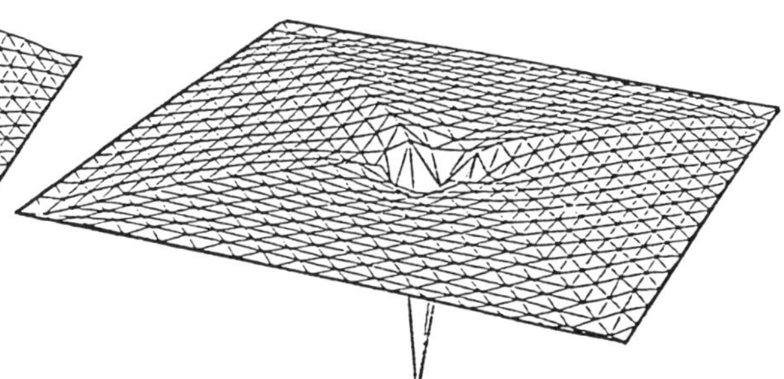

Fig. 2.5. 1st correction step (× 500).

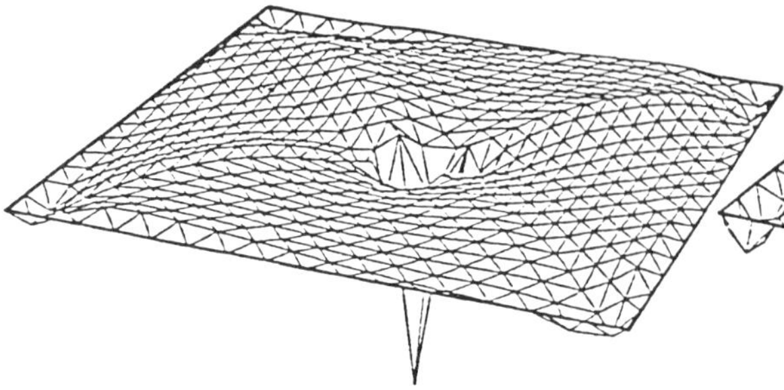

Fig. 2.6. 2nd correction step (× 2000).

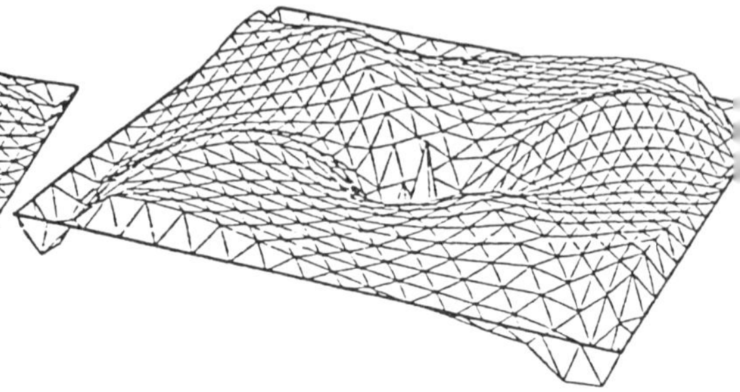

Fig. 2.7. 5th correction step (× 5000).

In many practical problems one is confronted with the presence of local "corner singularities" in the solution caused by reentrant corners or changes in the boundary conditions. These irregularities in the underlying differential operator induce a global "pollution effect", i.e., the overall accuracy of the discretization is drastically reduced even away from the corners. The use of extrapolation techniques in this situation is described in [3] and [19]; see also [23]. Also local defect correction appears to be possible for suppressing the pollution effect; see [12] for a discussion in connection with the multigrid method. In principle, the analysis presented above extends to the 3-dimensional case, with more or less obvious notational modifications. However, rigorous theoretical results and numerical calculations for practical problems of interest are rather rare. This field requires further investigation in the future.

3. Applications to Mixed Problems

The defect correction method is particularly attractive in the discretization of saddle point problems, since the higher order operator used in the defect computation does not need to be stable. We exemplify this through two numerical examples. The corresponding theory has been developed in [1] from where also the plots are taken.

The first example is the bi-harmonic problem

$$\Delta^2 u = f, \text{ in } \Omega, \quad u_{|\partial\Omega} = \partial_n u_{|\partial\Omega} = 0, \tag{3.1}$$

over the unit square $\Omega = (0,1)\times(0,1)$, which plays a central rool in plate bending theory and also for 2-d viscous flow problems. This problem may be written in the "mixed" formulation

$$(\nabla u,\nabla\psi) + (v,\psi) - (\nabla v,\nabla\varphi) = (f,\varphi), \quad \forall \{\varphi,\psi\}\in W_0^{1,2}(\Omega)\times W^{1,2}(\Omega), \tag{3.2}$$

where the unknowns are $u \in W_0^{1,2}(\Omega)$ and $v = \Delta u \in W^{1,2}(\Omega)$. The simplest mixed finite element scheme generates simultaneous approximations $\{u_h,v_h\}\in S_h\times\overline{S}_h$ to the pair $\{u,v\}$

through the saddle point equation

$$(3.2) \qquad (\nabla u_h, \nabla \psi_h) + (v_h, \psi_h) - (\nabla v_h, \nabla \varphi_h) = (f, \varphi_h), \quad \forall \{\varphi_h \psi_h\} \in S_h \times \bar{S}_h .$$

Here, $S_h \subset W_0^{1,2}(\Omega)$ and $\bar{S}_h \subset W^{1,2}(\Omega)$ are the spaces of linear finite elements, defined in Section 1, with and without zero boundary values, respectively. To this base discretization the defect correction is applied with (patchwise) bi-quartic interpolation for the variable u_h and bi-cubic interpolation for the variable v_h. The resulting errors against a polynomial reference solution are shown in the Fig. 3.1-3.4 , below. We note that in this case iterated defect correction further reduces the errors by more than one order of magnitude. Similar results have also been obtained for the discretization of problem (3.1) by the Herrmann-Miyoshi scheme which simultaneously approximates u and the full tensor of second derivatives $\nabla^2 u$; see [1].

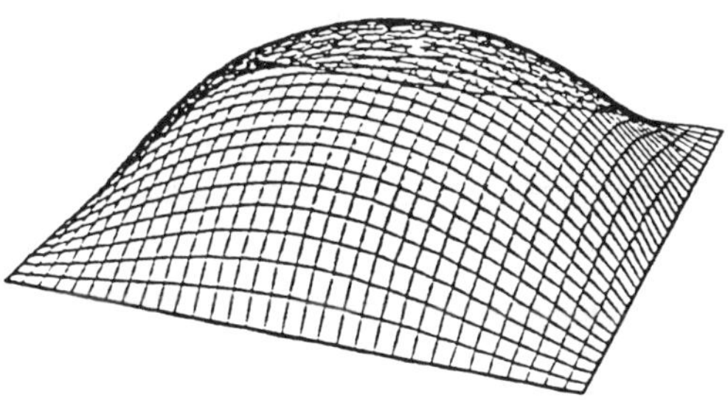

Fig. 3.1. Base error in u_h (× 60).

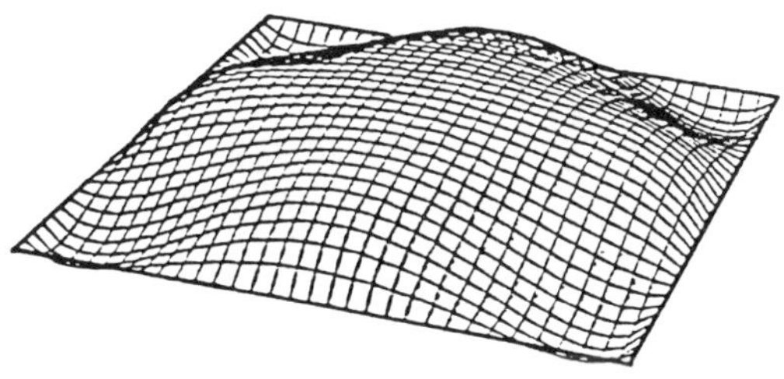

Fig. 3.2. Corrected error in $\bar{u}_h$ (× 6000).

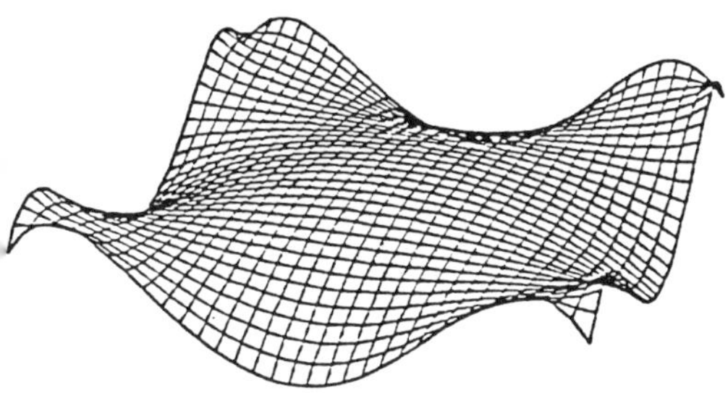

Fig. 3.3. Base error in v_h (× 1).

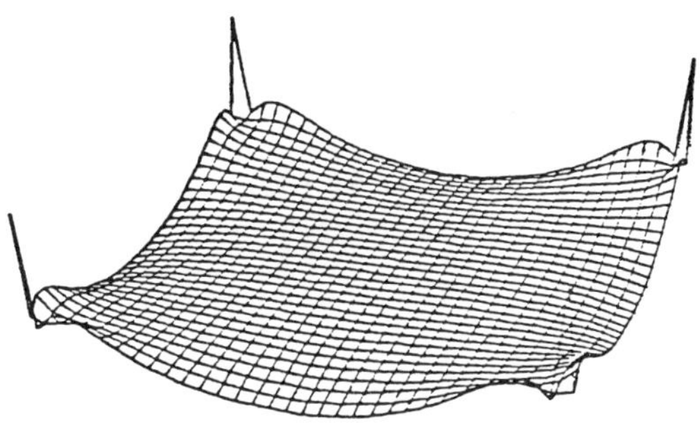

Fig. 3.4. Corrected error in $\bar{v}_h$ (× 20).

The second example is the Stokes problem

$$(3.2) \qquad -\Delta \vec{u} + \nabla p = \vec{f}, \quad \nabla \cdot \vec{u} = 0, \quad \text{in } \Omega, \quad \vec{u}_{|\partial\Omega} = 0,$$

again considered in the unit square $\Omega = (0,1)\times(0,1)$. Here, $\vec{u}$ and p are the velocity and the pressure, respectively, of a viscous incompressible flow. The discretization of this problem naturally leads to "mixed" formulations,

$$(3.3) \qquad (\nabla\vec{u}_h,\nabla\vec{\varphi}_h) + (p_h,\nabla\cdot\vec{\varphi}_h) - (\chi_h,\nabla\cdot\vec{u}_h) = (\vec{f},\vec{\varphi}_h)\,, \quad \forall\,\{\vec{\varphi}_h,\chi_h\}\in\vec{S}_h\times L_h\,,$$

in finite element spaces $\vec{S}_h\subset[W_0^{1,2}(\Omega)]^2$ and $L_h\subset L^2(\Omega)$. If the pairing $\{\vec{S}_h,L_h\}$ satisfies the Babuska-Brezzi stability condition

$$(3.4) \qquad \inf_{q_h\in L_h}\sup_{\vec{v}_h\in\vec{S}_h}\left\{\frac{(q_h,\mathrm{div}\,\vec{v}_h)}{\|q_h\|_{L_h/\mathbf{R}}\|\nabla\vec{v}_h\|}\right\} \ge \beta > 0\,,$$

then, the approximations $\{\vec{u}_h,p_h\}$ are uniquely determined in $\vec{S}_h\times L_h/\mathbf{R}$ and converge to $\{u,p\}$, with optimal orders. On a uniform rectangular mesh we take the space of piecewise bi-linear C^0-vector fields (vanishing along $\partial\Omega$) for $\vec{S}_h$, and that of piecewise constant functions for L_h . This pairing fails to satisfy the stability condition (3.4), but a patchwise filtering process generates a proper subspace $\hat{L}_h\subset L_h$, for which (3.4) holds true. The defects are calculated by using bi-cubic Hermite interpolation for the velocity and (discontinuous) bi-cubic Lagrange interpolation for the pressure. The reference solution is again taken to be polynomial.

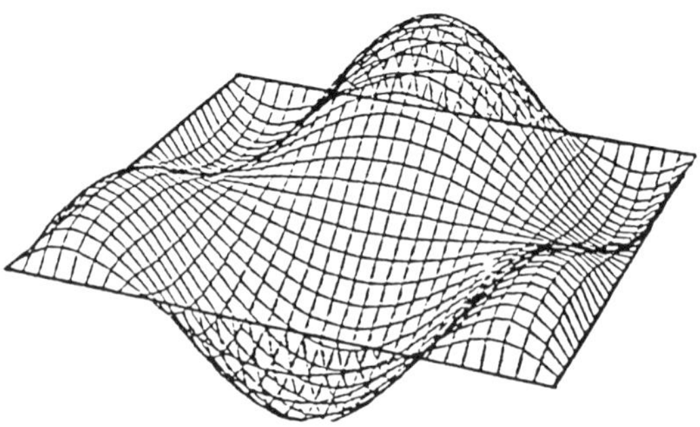

Fig. 3.5. Velocity error ($\times$ 100).

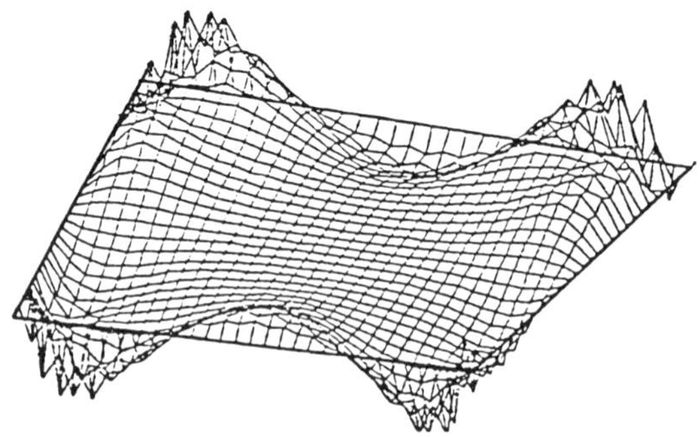

Fig. 3.6. Corrected velocity error ($\times$ 10000).

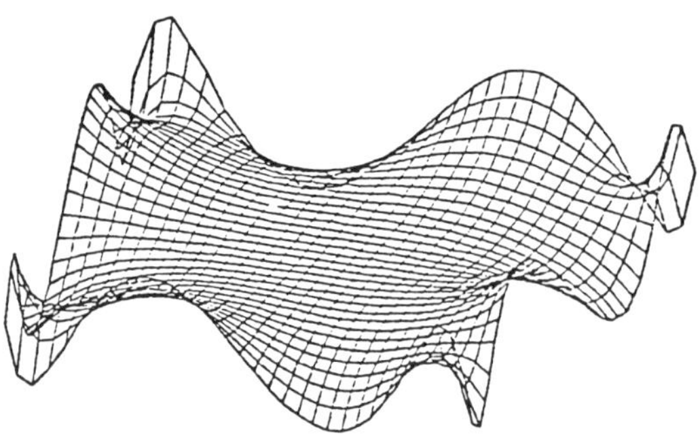

Fig. 3.7. Pressur error ($\times$ 8).

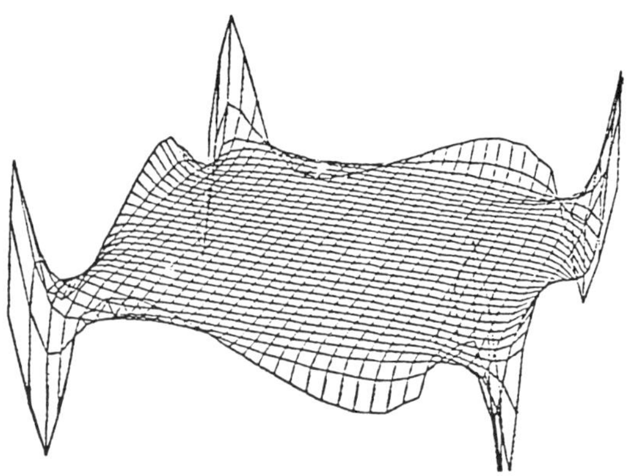

Fig. 3.8. Corrected pressure error ($\times$ 100).

4. Time Dependent Problems

We briefly discuss the application of the defect correction method to time dependent problems. To this end, let us consider the parabolic model problem

(4.1) $$u_t - \Delta u = f, \text{ in } \Omega\times(0,\infty), \quad u_{|\partial\Omega} = 0, \quad u_{|t=0} = u^0,$$

and its (spatially) semi-discrete counterpart

(4.2) $$u_{h,t} - \Delta_h u_h = f_h, \quad t>0, \quad u_h(0) = u_h^0,$$

where $f_h = P_h f$. The underlying finite element discretization is the same as that introduced in Section 1. In the case that the initial data u^0 is smooth (and compatible), one may take $u_h^0 = I_h u^0$, otherwise the appropriate choice is $u_h^0 = P_h u^0$. For the discrete solution $u_h(t) \in S_h$, there holds the L^∞-stability estimate (see [24])

(4.3) $$\sup_{[0,t]}\|u_h\|_\infty \le c_\varepsilon h^{-\varepsilon}\left\{\|u_h^0\|_\infty + \sup_{[0,t]}\|f_h\|_{-2,\infty;h}\right\}.$$

In the usual way one obtaines from (4.3) the quasi-optimal L^∞-error estimate

(4.4) $$\sup_{[0,t]}\|u-u_h\|_\infty \le c_\varepsilon h^{2-\varepsilon} \sup_{[0,t]}\|u\|_{2,\infty}.$$

Furthermore, on a three-directional mesh, one has the asymptotic expansion

(4.5) $$u_h(t) = I_h u(t) + h^2 I_h e(u;t) + O(h^{2+\kappa}).$$

with a certain function $e(u;t)$. Depending on the regularity of the problem (4.1), one has $\kappa \in (0,2)$. As in the elliptic case, the proof of these results are based on sharp L^1-estimates for discrete fundamental solutions which are derived using weighted Sobolev norms.

Again one may use the several defect correction approaches described in Section 1 in order to increase the accuracy of the base discretization (4.2). Let, for example $\tilde{I}_h$ be the operator of the cubic Lagrange interpolation, and let $\bar{u}_h^0 \in S_h$ be a higher order approximation to the initial value u^0, satisfying

(4.7) $$\|\bar{u}_h^0 - I_h u^0\|_\infty = O(h^{2+\kappa}).$$

Let $\bar{u}_h(t) \in S_h$, with $\bar{u}_h(0) = \bar{u}_h^0$, be the solution of the corrected equation

(4.8) $$\bar{u}_{h,t} - \Delta_h \bar{u}_h = f_h - \left\{P_h \tilde{I}_h u_{h,t} - \Delta_h R_h \tilde{I}_h u_h - f_h\right\}, \quad t>0.$$

Then, if $u(t)$ is sufficiently smooth, there holds

(4.9) $$\bar{u}_h(t) = I_h u(t) + O(h^{2+\kappa}), \quad t>0.$$

The argument is similar to that used in Section 1, for the elliptic case, now basing on the error expansion (4.5) and on the stability estimate (4.3). Again, observing that $\tilde{I}_h I_h = \tilde{I}_h$, a simple calculation yields

$$(\bar{u}_h - I_h u)_t - \Delta_h(\bar{u}_h - I_h u) = P_h w_t - \Delta_h R_h w ,$$

where $w \equiv (1-\tilde{I}_h)(u_h - I_h u) + u - \tilde{I}_h u$. Consequently, by (4.3), it follows that

$$\sup_{[0,t]}\|\bar{u}_h - I_h u\|_\infty \le c_\varepsilon h^{-\varepsilon}\Big\{\sup_{[0,t]}\|(1-\tilde{I}_h)(u_h - I_h u)\|_\infty + \sup_{[0,t]}\|u - \tilde{I}_h u\|_\infty\Big\} .$$

In view of the expansion (4.5), the error $u_h - I_h u$ is "smooth", and there holds

$$(1-\tilde{I}_h)(u_h - I_h u) = h^2(1-\tilde{I}_h)e(u) + O(h^{2+\kappa}) .$$

Observing that $\|(1-\tilde{I}_h)e(u)\|_\infty = O(h^\kappa)$, the error estimate (1.15) completes the proof of (4.9). In the case of a convex polygonal domain, the foregoing results remain valid on certain subdomains $\Omega' \subset\subset \Omega$.

In order to actually compute the semi-discrete approximation $u_h(t)$, one has to employ a time stepping scheme. As an example we consider the popular second-order Crank-Nicolson scheme. Let k denote the time step (for simplicity constant) and $t_n = nk$. Then, the fully discrete solutions $U^n_{h,k}$ are determined through the relations $U^0_{h,k} = u^0_h$, and

(4.10) $$(1-\tfrac{k}{2}\Delta_h)U^n_{h,k} = (1+\tfrac{k}{2}\Delta_h)U^{n-1}_{h,k} + kf_h^{n-1/2} , \quad n \ge 1 ,$$

where $f_h^{n-1/2} \equiv f_h(t_{n-1/2})$. For this, there holds the L^∞-error estimate (see [25], [24])

(4.11) $$\|U^n_{h,k} - u(t_n)\|_\infty \le c_\varepsilon(h^{2-\varepsilon} + k^2) ,$$

provided that the solution u is sufficiently smooth. The principle of defect correction may be applied to the fully discrete scheme (4.10) simultaneously with respect to space and time. However, it appears more natural to use the defect correction only in space and (global) Richardson k-extrapolation in time; see [26]. The theoretical basis of this approach are an asymptotic expansion of the form

(4.12) $$U^n_{h,k} = I_h u(t_n) + h^2 I_h e(u;t_n) + k^2 I_h d(u;t_n) + O(h^{2+\kappa} + k^{2+\lambda}) ,$$

with certain functions $e(u;t)$ and $d(u;t)$, and a stability estimate of the form

(4.13) $$\max_{0\le\nu\le n}\|U^\nu_{h,k}\|_\infty \le c_\varepsilon h^{-\varepsilon}\Big\{\|U^0_{h,k}\|_\infty + \max_{0\le\nu\le n}|||f^\nu_h|||_{-2,\infty;h}\Big\} .$$

Expansions of the form (4.12) are known for certain simple finite difference schemes on uniform rectangular grids (see [21]). For finite element discretizations corresponding results are very likely to hold, in view of the existing analysis for the elliptic case. The stability estimate (4.13) can be proven by the techniques developed in [24] for the backward Euler scheme.

References

1. Blum,H.: Asymptotic error expansion and defect correction in the finite element method, Habilitationsschrift, Universität Heidelberg 1990.
2. Blum,H., Lin,Q., Rannacher,R.: Asymptotic error expansion and Richardson extrapolation for linear finite elements, Numer.Math. 49, 11-37 (1986).
3. Blum,H., Rannacher,R.: Extrapolation techniques for reducing the pollution effect of reentrant corners in the finite element method, Numer.Math. 52, 539-564 (1988).
4. Blum,H., Rannacher, R.: Finite element eigenvalue computation on domains with reentrant corners using Richardson extrapolation. J.Comput.Math. 8 (1990), to appear.
5. Böhmer,K., Hemker,P.W., Stetter,H.J.: The defect correction approach, in "Defect Correction Methods. Theory and Applications" (K.Böhmer, H.J.Stetter, eds.), pp. 1-32, Computing Suppl. 5, Springer 1984.
6. Ciarlet,P.G.: The Finite Element Method for Elliptic Problems. North-Holland 1978.
7. Dobrowolski,M., Rannacher,R.: Finite element methods for nonlinear elliptic systems of second order, Math.Nachr. 94, 155-172 (1980).
8. Fairweather,G., Saylor,A.: On the application of extrapolation, deferred correction and defect correction to discrete-time Galerkin methods for parabolic problems, IMA J.Num. Analysis 3, 173-192 (1983).
9. Frank,R., Hertling,J., Monnet,J.P.: The application of iterated defect correction to variational methods for elliptic boundary value problems, Computing 30, 121-135 (1983).
10. Frehse,J., Rannacher, R.: Eine L^1-Fehlerabschätzung für diskrete Grundlösungen in der Methode der finiten Elemente. Tagungsband "Finite Elemente", Bonn.Math.Schr. 89, 92-114 (1976).
11. Greus,M.: Defektkorrektur-Methoden für lineare finite Elemente bei der Poisson-Gleichung, Diplomarbeit, Universität Saarbrücken 1989.
12. Hackbusch,W.: Local defect correction method and domain decomposition techniques, in "Defect Correction Methods. Theory and Applications" (K.Böhmer, H.J.Stetter, eds.), pp. 89-114, Computing Supplementum 5, Springer 1984.
13. Hanson,P.M., Walsh,J.E.: Asymptotic theory of the global error and some techniques of error estimation, Numer.Math. 45, 51-74 (1984).
14. Lin,Q., Lü,T., Shen,S.M.: Maximum norm estimate, extrapolation and optimal points of stresses for the finite element method on strongly regular triangulation, J.Comput.Math. 1, 376-383 (1983).
15. Lin,Q., Zhu, Qi-Ding: Local asymptotic expansion and extrapolation for finite elements, J.Comp.Math. 4, 263-265 (1986).

16. Lin,Q., Xie,R.F.: How to recover the convergence rate for Richardson extrapolation on bounded domains, J.Comput.Math. 6, 68-79 (1988).

17. Lin,Q., Xie,R.F.: Error expansions for finite element approximation and its applications, Proc. First China Conf. Numerical Methods for Partial Differential Equations, Shanghai, April 1987 (You-Lan Zhu, Ben-Yu Guo, eds.), Springer 1987.

18. Lin,Q., Xie,R.F.: Some advances in the study of error expansion for finite elements. I. Eigenvalue error expansion, J.Comput.Math. 4, 368-382 (1986).

19. Lin,Q.: Fourth order eigenvalue approximation by extrapolation on domains with reentrant corners, Preprint No. 551, SFB 123 Universität Heidelberg, December 1989.

20. Lindberg,B.: Error estimation and iterative improvement for discretization algorithms, BIT 20, 486-500 (1980).

21. Marchuk,G., Shaidurov,V.: Difference Methods and their Extrapolations, Springer 1983.

22. Rannacher,R.: Richardson extrapolation with finite elements, Proc. 2nd GAMM-Seminar "Numerical Techniques in Continuum Mechanics", Kiel, 1986 (W. Hackbusch, C.Witsch, eds.), pp. 90-101, Vieweg 1987.

23. Rannacher,R.: Extrapolation techniques in the finite element method (A survey), Proc. "Summer school on Numerical Analysis", Helsinki 1987 (O.Nevanlinna, ed.), pp. 80-113, Helsinki, University of Technology, Report-MAT-C7 (1988).

24. Rannacher,R.: L^∞-stability estimates and asymptotic error expansions for parabolic finite element equations, Proc. GAMM-Seminar "Extrapolations- und Defektkorrektur-methoden", Heidelberg, June 22-23, 1990 (J.Frehse, R.Rannacher, eds.), to appear

25. Schatz,A.H., Thomeé,V., Wahlbin,L.B.: Maximum norm stability and error estimates in parabolic finite element equations, Comm.Pure Appl.Math. 33, 265-304 (1980).

26. Seward,W.L., Fairweather,G., Johnston,R.L.: A survey of higher-order methods for the numerical integration of semidiscrete parabolic problems, IMA J.Numer.Analysis 4, 375-425 (1984).

27. Skeel,R.D.: A theoretical framework for proving accuracy results for deferred corrections, SIAM J.Numer.Anal. 19, 171-196 (1981).

28. Stetter,H.J.: The defect correction principle and discretization methods, Numer.Math. 29, 425-443 (1978)